TEACH YOUR
CHILDREN WELL

TEACH YOUR
CHILDREN WELL

CHOON TAN
with
Veronika Meduna

CANTERBURY UNIVERSITY PRESS

First published in 1996 by
CANTERBURY UNIVERSITY PRESS
University of Canterbury
Private Bag 4800
Christchurch
New Zealand

ISBN 0-908812-51-5

Designed and typeset by Richard King
Printed by GP Print, Wellington

Cover design by G. & A. Nelson
Cover photograph by Rob Tucker

Contents

Introduction – the Tan family

MORE THAN TWENTY YEARS AGO, WHEN I BEGAN teaching mathematics to my seven-year-old son, David, I had no idea that he would later become the youngest university graduate in New Zealand.

I had even less of an idea that all our three children would go through their schooling much faster than expected, leaving schoolteachers, academics and psychologists amazed by their excellence in maths, their quick and sharp minds and their keen fascination with everything.

Although I am undeniably a very proud father and happy for our children, none of their academic achievements would have any value to me if I couldn't see that their lives continued to be rich and fulfilling beyond childhood and that each of them has grown into a self-assured, balanced and confident young person.

Our youngest son, Michael, is only eleven and the only one of our children living with us at the moment. His academic achievements have been outstanding as he passed his bursary maths exam at the age of seven, but he is also a very kind young child, always happy to help and very popular with his peers – friends in his own age group as well as his fellow university students. We continue learning together at home while he attends maths lectures at university and, apart from a few occasions, I cannot remember him feeling upset, angry or frustrated.

Our daughter, Audrey, is studying for her PhD in pure mathematics at Cambridge University. She seems to live the life of any twenty-year-old with a large group of friends, including some very special friendships, and has many interests outside her university work. As a teenager she had spent a few years putting everything into music and ballet, and she still loves these activities. Her friends and lecturers have

7

described her as a very outgoing, open-minded and friendly young woman.

Our eldest son, David, who will soon be twenty-nine, is working on a postdoctoral project developing mathematical models with a group of researchers looking into the depletion of the ozone layer. He is a modest young man and has a confidently relaxed sort of nature. I remember how he was embarrassed by all the media attention he received when he got his first university degree at sixteen.

Teaching our own children, I discovered how satisfying teaching could be for myself. Together with my wife, Rosemary, I began teaching others, and over the years we have coached hundreds of children and teenagers through their school exams in maths and, perhaps more importantly, through their difficulties with the subject. Even though our teaching has always been focused on maths, we knew intuitively that learning progresses best if a child is not distracted by negative attitudes, lack of motivation or zeal for learning, put-downs and feelings of inadequacy or anxiety about mathematics.

Over the years I developed Rapid Mathematics, a method that basically encapsulates my main goal of covering the mathematics syllabus as quickly and efficiently as possible with the highest possible degree of choice and control for the child.

All we teach is maths, but our teaching method attempts to focus on each individual, trying to teach the child rather than the material. Apart from tackling particular problems in maths, we help children to change their attitude towards the subject and learning generally, and we return to them their confidence in their own skills and capability. They regain control over their progress relaxed in the knowledge that there is somebody to help over any present hurdles and that they will learn how to overcome any future barriers themselves.

While we tried to help children to understand and cope better with maths, we have learned a lot ourselves about things I believe are essential for parents and teachers to know if we want to teach our children well and provide them with the best possible start in their lives. This book is about these ideas and experiences, and about children's potential to grow into self-assured and capable young people. This growth is open to all children because another important thing we have learned is that our own children are not really the exception they were often portrayed to be: that children all have a vast potential to achieve – a potential just waiting to be tapped into.

Nobody is born extra smart!

THE 'NATURE VERSUS NURTURE' ARGUMENT IN education is centuries old and has never failed to spark heated debate in the scientific community or among teachers and parents. Children and adults who are achieving beyond the expected and accepted boundaries of learning have always attracted plenty of admiration, fascination and respect, but also controversy, envy and plain rejection.

My own family was often in the midst of it all. Our three children all sat University Entrance exams in maths while still at primary school, and were among the world's youngest university entrants. We received a lot of media attention each time any of our children did something that hadn't been achieved before. The descriptions ranged from a crazed family hot-housing their children, whose genetic and cultural inheritances allowed them to be pushed far beyond other children, to a never-to-be-repeated, freak accident of nature producing three maths prodigies in one family. This was the vantage point preferred by most who made comments about us especially when Michael, our youngest, came along to outdo his siblings on the academic scale. He passed his Bursary maths exams at seven, which earned him an entry in the *Guinness Book of Records* in 1992.

However, those who made the effort to meet us often admitted to being pleasantly surprised about the relaxed easygoing atmosphere at our home and the children's numerous other activities and interests which they followed with the same enthusiasm they have for maths. It is their attitude towards their learning that makes them perform beyond the normally expected levels. None of them is genetically advantaged, born with some sort of gift or talent; they are not a breed apart from others. In fact, after years of teaching other children, many of whom were considered failures in maths, I am convinced that there is no such thing as a genius or a gifted person but that it is open to all of us, including the so-called non-achievers, to get closer to our maximum potential if we manage to change our attitude towards learning and achievement. And once there, any intellectual activity will be so much more rewarding and motivating that learning will become the pleasure it should always be.

Although we are all constantly learning throughout our lives, children's learning during their school years is more structured and their performance and achievements at school often become some of the most important things in their lives. Because they spend most of their time in a learning situation, their progress takes the same priority as work-related issues for adults. For this reason, and because the childhood and teenage years have such an influence on a person's future character and personality, parents, teachers and everybody in contact with children should always be aware of the importance of an encouraging and supporting approach.

Maths might sound an unlikely subject to help children on their way to more efficient and enjoyable learning in general, but it is the perfect topic to start off their analytical and creative skills and to keep up or boost their naturally inquisitive nature. And if children are taught maths early and well, they have a good chance of bypassing the disabling difficulties the subject can often cause for secondary school children. They will also develop confidence and knowledge about their particular strengths and weaknesses, enabling them to approach any other subject or future problem secure in the knowledge that they will be able to achieve whatever is required.

This feeling of security is exactly what we attempt to provide for our own children and students who come for tutoring. We want the children to develop their potential and enjoy life's opportunities and challenges to the fullest. And we are not the only family in history to attempt and succeed at that. There are other families with children who have shown signs of excellence at an early age when neither parent had any obvious brilliance. For me, this shows that children's genetic makeup is only of secondary importance.

One example is the Polgar family from Hungary, whose three daughters gave the competitive-chess world a considerable shake-up when they beat some of the current (mostly male) champions while still in their teens. All three were home-taught by their parents with a strong focus on chess and developed an excellence and mastery of the game, which surpassed most other players. The girls were taught chess from as young as three years, but they showed the fastest progress only after chess had become a genuine passion for them and they had moved from being taught to learning actively from any source available to them.

This is a crucial point. If we manage to help children to take control over their learning instead of passively consuming what teachers put in front of them, they will progress a lot faster, with more confidence and with a lot more fun.

Each of the Polgar sisters managed to maintain their love for chess as young women, and they continue to be successful in competitive circles. Although I would agree that chess will provide a child with many useful skills such as sharpening their mind and teaching them to plan ahead, analyse somebody else's tactics and co-ordinate the many different possible moves, I would not focus on the game as strongly as the Polgars. Being strong competitive chess players will certainly open many doors, but to me it seems too unrelated to many everyday tasks to justify the early, strict specialisation.

Clearly, there are also examples that could be classified as early versions of hot-housing. The Stern family made headlines in the United States in the 1970s when father Aaron Stern held a press conference to announce to the world the birth of his genius daughter Edith. He was convinced he could educate his children and mould them into 'geniuses'. Edith and her younger brother David were constantly exposed to classical music even if it was only as a background to other activities and, from infancy, their days were filled with a strict schedule of educational activities. The father has been described as loving but determined and single-minded.

One can easily anticipate how such an approach, although successful as far as the extraordinary academic achievements of the children are concerned, could prepare the scenario for later conflict and unhappiness. Any parent will know how difficult it can be to let children grow and develop as unique individuals without any restrictions from parents' ideals and expectations. But if children are to grow into confident and competent adults they have to have the opportunity and the freedom to be themselves no matter how far away from the parents' dream they are.

One of the most heart-warming stories showing that the right support and encouragement can have more influence than children's genetic outfit in creating abilities at a young age is the story of the La Pieta orphanages in Venice, which I read in Michael Howe's *Sense and Nonsense about Hothousing Children*. At the beginning of the eighteenth century, Italian composer Antonio Vivaldi worked at the orphanage teaching the girls how to play the violin. Music was very highly valued at the orphanage and the children were strongly encouraged and supported in their efforts to become accomplished performers and singers. Many children reached levels of performance that attracted visitors from far away.

The girls' performances were described as unsurpassed in the world, despite the fact that many of them came from backgrounds of poverty, destitution and deprivation. The orphanage carried out an unplanned 'natural experiment' which seems to contradict the view that outstanding skills, in this case musical talents, can be gained only by those fortunate enough to be born with the gift.

Many of my students have showed me clearly that there doesn't need to be a talent or gift to achieve above-average or very high standards. Even motivation or the lack of it is not a real problem because most people are keen to move on and continue adding to their knowledge and experience as long as they find a learning environment which allows them to do it without having to waste energy on comparison and competition with others, or constant efforts to maintain their feeling of self-worth. Many children seem to be held back because they are unable to find the trust in their own abilities, confidence that they will be able to achieve whatever their goal is, and an ongoing supply of encouragement and support.

I am happy to leave the academic debate about the exact relationship between nature and nurture to the education experts, but I would like to share some of my thoughts about what to do if we want to teach our children well. I am sure readers will recognise many of the ideas and thoughts as their own and common sense. They will also know the difficulty of always carrying out exactly what you think is the best way of approaching any particular problem. But there is no reason to panic or feel disheartened and frustrated even if we sometimes fail to do what we know is our best. Intentions are often more important than the actual achievement, and intuition and spontaneity make any relationship between people more genuine and credible. And in any human relationship there has to be space to allow for the individuality of each person.

With this book, I would like to establish a flexible guide rather than a set of rigid rules, because each child is too much of an individual to be approached relying on a set of general principles.

Why maths?

MATHS CAN TROUBLE AND DEMOTIVATE CHILDREN for years. It frustrates me each time I meet a new student to realise how many children don't enjoy school because they find maths difficult. Their fears are often not even based on their own experience or failures but on their older peers' problems. Maths and maths-related subjects generally have a reputation for being boring, too hard, too complicated, only for bright people and even strange and useless. Yet, mathematics could and should be *the* most rewarding and enjoyable subject.

Many adults also associate maths and related subjects

with a sense of dusty and somewhat sterile irrelevance. Just mentioning words like fractions, trigonometry, or equations makes many children and adults feel uncomfortable and takes away whatever motivation for learning they might have.

I experienced the crippling effect of fear of maths first-hand when I was at high school, a boarding school with three hundred boys. The older boys made us feel petrified at the prospect of having to learn algebra. They all complained to us about how horrible and complicated it was, and years before we even started learning anything about algebra ourselves we were already afraid of failure. This feeling has probably affected many children at some stage during their schooling or even before they start school, disregarding their actual aptitude for maths and the fact that the subject can be presented in a way that gives every child a chance to achieve. All my teaching focuses on freeing children from these strains, fears and hangups.

However, when first starting to teach our own children, I chose maths mainly because it was a subject I felt confident with, and I knew it was often neglected during the early years of a child's development. I used to help my own classmates with mathematical problems at high school and I knew I could get pleasure and satisfaction out of helping others. I was always motivated to keep learning myself but my teaching was not based on any particular method. I was following my intuition more than principles.

The teaching sessions with our children began quite accidentally as a way of providing entertainment for David following the birth of his sister Audrey, who was receiving a lot of our attention. He sometimes felt a bit lost and often asked me to play chess with him. I am not very good at chess and wasn't keen on the competitive side of the game. In order to work in a team together rather than compete against each other I suggested working together on his school maths with him and we began working through his book.

Only much later, after teaching many other children, did I conclude that mathematics is the perfect subject to provide basic learning and problem-solving skills as well as confidence to tackle other subjects, because maths incorporates creative as well as logical and structural thinking. It is an intrinsic, intuitive skill, easily measurable and requires no body of prior experience or knowledge.

Rosemary was more interested in the arts, languages and history during her school and university years and thought she had no aptitude for maths and the sciences. For a long time she felt these options had closed for her when she left school. It was only when David prepared for his University Entrance maths exams at night classes that she decided to join him. The classes were mainly for adults who intended either to brush up on their skills or to get their University Entrance certificate for a new career. Each evening, a few children from families who couldn't find a babysitter would sit in the back of the room reading or playing. The supervisor of the night classes was delighted and amused to see the somewhat reversed roles of David and Rosemary! He was later to become one of David's supporters when our son started thinking about enrolling in university maths lectures as a ten-year-old.

Rosemary attended classes with all our children, mainly because she wanted to accompany them as they were so young. But by actually sitting the exam, joining Michael at university lectures and assisting me with the teaching of other children at our home, she began to enjoy the subject and overcame her antipathy against things mathematical.

She often said to me that her experience was probably typical. She believes many mothers and perhaps even primary teachers, a profession often dominated by women, might feel just as inadequate as she initially did. While many families would consider taking their children to libraries and reading books with them as a natural part of their upbringing, they hardly ever consider maths. Many parents don't feel confident or competent enough, so many children encounter mathematics for the first time only in pre-school or school. Yet mathematical operations are hidden in many everyday activities: shopping, cooking for the family, planning a holiday trip, setting the alarm clock, and an endless list of other things which we do every day without realising that they could be used to introduce children to maths.

Mathematical competence is to education what physical fitness is to sport: it trains the brain for the challenges of life; it teaches to think logically, to reason, to be flexible and imaginative. Unlike any other subject, maths combines skills needed in most other subjects. If taught well, it can cover many important principles far beyond arithmetic and numeric skills. Science tends to need mathematics more; languages and arts have a stronger focus on the creative skills. In contrast to other subjects, maths is very objective and a bit like a puzzle – there is always a clear solution, which almost guarantees a very satisfying feeling of achievement.

Because it is almost like a separate language replacing commands with symbols which can be built into rather sophisticated sequences just like sentences, maths teaches all the skills needed to express things in abstract ways and to create generalisations and rules. Maths also teaches problem-solving skills, because there are always several ways of arriving at the right solution but in most cases only one right answer. It can be quite motivating to know that, although it might seem complicated, there will be a way to solve the puzzle.

Used as a basic tool in all sciences, maths teaches analytical thinking and understanding of graphical explanations and displays, something which has become increasingly important in many jobs. Issues of fairness, integrity and justice, even moral questions can easily find their way into a maths lesson. For example, a lesson about division can incorporate issues of fairness and further discussion can help the child to develop their own moral principles. The following dialogue could happen in any maths class or in any home.

The maths question is: 'How many pieces of cake would you get in a family of four (Mum, Dad, little sister and you) if the cake had been cut into eight pieces?'

The mathematically correct answer would be two pieces each, but the examples could be extended to see what happens if Dad is concerned about putting on more weight? 'Two more for me!' might be the spontaneous answer, and suggestions that Mum had been advised by her dentist not to eat too many sweets could follow, but soon the whole group would be discussing how to share any extra cake with other members of the family. Any other exercise can be extended to include other issues and to help children to enjoy thinking through mathematical problems. A mathematically competent student is likely to improve and become more competent and confident in many other areas of study, just as a physically fit person is likely to excel in many sports. Most of my students have not only made faster progress in maths but have showed across-the-board improvements in many subjects once they have gained enough confidence and begun to change from expecting failure to believing in success and accomplishment.

Why maths? Because . . .

✔ Maths incorporates creative as well as logical and structural thinking.
✔ If taught well, maths can be great fun.
✔ Maths is everywhere: many everyday activities include a bit of maths.
✔ There's a lot of discovery in maths.
✔ It's the basis for many subjects and teaches skills useful in most other subjects.
✔ It does not require any previous knowledge.
✔ It's often neglected during early childhood years.
✔ It makes school easier.
✔ It does not need to be difficult and complicated.
✔ It rewards through a strong sense of achievement and accomplishment.
✔ It's a good way to start learning problem-solving skills.
✔ It teaches confidence to tackle and solve other tasks.
✔ Teaching maths can incorporate issues of fairness and integrity and moral questions.

What we see in reality, however, is a large number of students failing in maths and, as a consequence, they are more susceptible to low self-esteem and a negative attitude towards their schooling and learning in general. Teachers are often at their wits' end and don't know how to motivate the student to do more. They are often under pressure to progress through the syllabus at a reasonable rate and don't seem to have the time to give each child individual attention. Parents often feel helpless because they are not particularly comfortable with maths either, or because their attempts to help are rejected by their children.

My child doesn't listen to me . . .

WHAT CAN PARENTS DO WHO WANT TO HELP THEIR children in maths but can't make them listen or concentrate and are told that the teacher knows everything better and teaches them different things?

In families where parents have left their children's education entirely to the teachers and only attempt to help when the children develop problems, it can often be difficult to get the necessary amount of respect and rapport from the children, especially if they are of secondary school age. Children can find it impossible to appreciate their parents' help and regard it as interference rather than coaching.

I often encountered a similar reaction with my students, particularly those who have been talked into taking up maths tutoring by their family. They resent the whole idea of having to do more in a subject they don't like and feel they have failed with. They feel worse because they think they have disappointed their parents who want them to perform better. Caught somewhere between the fear of further failure and the desire to please their parents, they are often quite tense and not very energetic. They are already under pressure and the last thing they want is somebody else who tells them what to do and how to do it.

One way to deal with this sort of situation is to let the child be in control. For example:

'Can you show me what sort of exercise you last did in maths?'

Your child might still be reluctant but will at least show you the page.

'How would you get started on these exercises?'

Your child might start explaining how to do them or just say, 'I can't do it.'

'Let's go back to something not as advanced and see if you find that easier to do. Perhaps we can find out where the difficulty is.'

(If it is the first time you look into your child's maths textbook you might feel genuinely impressed by the level of difficulty – and it doesn't hurt to mention that.)

'Well, I don't know if I could do that but can you explain what you find difficult?'

'I just can't do it.'

'Let's go through it step by step together, shall we? Perhaps we can work it out together.'

Anything that involves children in a dialogue instead of exposing them to a lecture will give them a feeling of control and make them more interested in participating. They have to know that they can always tell you what they think and feel.

Often they will give you an insight into many other aspects of their life which could well influence their performance, too. They might mention that they feel inferior because somebody in class has been bullying them; that their teacher seems to think they can't do maths anyway; that their best friend thinks maths is not an important subject; or that they have difficulty hearing the teacher properly or reading the blackboard.

Although you might want to focus on getting the maths problem out of the way, the child's problems are likely to be caused by many factors, some of which could also be physical, such as eyesight or hearing problems, and might need specialist help.

If the child continues rejecting your help, you might be able to swap roles entirely and let the child be the teacher. This happened with David when I began teaching him maths at the age of seven. I tried to do one of the subtractions in his text book:

$$\overset{\overset{3}{}\overset{1}{}}{\cancel{4}3}$$
$$-\ 29$$
$$\overline{\ 14\ }$$

The way I had learned subtractions was this. Firstly, there is not enough in 3 to subtract 9, so you have to 'borrow 1 from the 4' in the top line to get 13. Then, subtracting 9 from 13 gives you 4. This leaves you with only 3 in the top line where the 4 was; from this you subtract the 2 in 29 to get the 1 in the bottom line.

David, however, had been taught to borrow the 1 from the lower line and then 'bring it back' and add it to the 2. He rebuffed me, saying that his teacher had used other methods to do the exercise.

He had been taught to do the same exercise like this:

$$\overset{1}{}$$
$$43$$
$$-\ \overset{1}{2}9$$
$$\overline{\ 14\ }$$

Of course we both got the same answer, and he had to acknowledge that there seemed to be several ways to do the same thing. When I suggested that he could teach me what he knew and what he had learned from his teacher, he enjoyed the session so much that he gradually allowed me to take over and introduce new things.

Years later, after David had passed his University Entrance maths exams at the age of ten, he was helping me teach other children. He had not only learned his maths a lot faster but also discovered a joy in teaching and had developed impressive skills in explaining maths and motivating others.

This trick has helped me with many other children who became very motivated as soon as they could be in control and felt they could explain something they knew to somebody else. I think many parents would be able to think back to a similar incident when they were children themselves.

Activities children might regard as unpleasant chores can become quite enjoyable if they feel they are not doing them only because they have to or because somebody expects them to, but because they can be helpful and somebody appreciates their skill and knowledge.

Not only will the role change motivate children: they will also automatically practise skills such as negotiating a point of view, checking whether what they suggest makes sense and comparing their own method with other options – all of which are important skills that are easily transferred into many other areas of study and life.

Some children will want to come back to the role games every now and then and they should be encouraged because it will not only help their own knowledge to settle more permanently in their memory, but it will also teach them negotiating skills.

One day, I had asked David to fill in for me during part of a tutoring session and he gave me an ear-to-ear grin and said, 'Hand over your red pen, Dad!' I had been using a red pen more by habit than as a symbol of authority. Although the red pen had no important meaning for me, I knew that he associated it with my scribbling in his exercise book and overseeing his work. I was more than happy to let him have it, as it only reinforced his enthusiasm for both the subject he was teaching and his joy in helping others who still struggled with maths. I always wanted my and other children to feel equal because the teacher-pupil relationship can easily deteriorate and create an atmosphere of pressure. This little gesture gave David an enormous boost at the time.

However, parents who teach their children earlier than we started with David will probably never have problems of acceptance, as long as they adopt an informal method of offering information to the child. Ordinary day-to-day activities like shopping or preparing food for the family can include small tasks that introduce ideas about numbers, quantity, space and geometric shapes.

As long as the relationship between parents and child is without major tensions, I believe most children will soak up any new material like a sponge.

How do children learn?

EDUCATIONISTS OFTEN COMPARE YOUNG CHILDREN who are confronted with new information, with a visitor to an unfamiliar but very exciting country. Seen for the first time, the new land appears stimulating and richly interesting, but at the same time deeply upsetting, sometimes overwhelming. I agree with this description but would extend it to apply throughout a person's life.

Many educational scientists have said that children are capable of learning more and faster than adults, and I agree that the pace and accuracy of learning seems to change with the years, but I don't think the overall capability and motivation fades naturally with age. Given an encouraging and supportive environment, teenage children and adults can rediscover their hunger for learning, their motivation to explore and their desire to achieve. I am often surprised by the energy some teenagers or older students develop when

Informalised education is

✔ Sharing your knowledge rather than lecturing children.
✔ Letting them show you what they know and how their teachers taught them to do things.
✔ Telling them if you're impressed with what they are doing.
✔ Making a conscious effort to talk to them as equals.
✔ Asking open questions.
✔ Letting them be in control.
✔ Being patient enough to listen to things that are not related to maths.
✔ Encouraging children to talk about any other problems they might have.
✔ Using everyday situations to demonstrate the relevance of maths.

finally given some guidance and they succeed at their task, whether it is maths or any other challenge. Many have spent years being pigeonholed as non-achievers, and rediscovering their skills and potential makes their drive and energy virtually explode. At the same time I am sad to see how much of their own motivation, curiosity and desire to learn has been scooped out of some by a lack of either sustained encouragement or direction and vision.

One of my teenaged students, who had been receiving counselling for a while because of his disruptive behaviour, surprised his parents, the psychologists and himself by becoming keen and motivated and a helpful young person. I was unaware of his counselling until his mother rang me to say that his feeling of self-worth and attitude towards learning had been changing with each time he managed to take the next mathematical hurdle.

Self-esteem and confidence are rather fragile qualities, and it doesn't take much to convince young children that they will fail. It is also easy to make them believe there is nothing to strive for. But the same qualities are also naturally available to a young child. Babies are born with a strongly developed self-awareness; in fact they spend more time exploring themselves than anything else at the start of their lives. Their naturally confident way of demanding our attention can be unnerving for parents but shows that young children don't know much about self-doubt and insecurity. I don't believe in letting children cry until they learn that they can't always have things their way. By doing so, they might well learn that but the seeds of insecurity and later hangups will have been sown, too.

We were young parents, but we soon and very happily adjusted to the new task. During the first years of our children's life, especially in David's case, Rosemary was the one who set the foundations for their future learning. She has a strong sense of fairness and simply a love for children. She enjoys watching them develop and grow, and she always dropped whatever she was doing when they wanted her attention. When we had David, we used to live upstairs and had a big, open balcony. By the time he was crawling, he certainly received more attention than the average child simply because of our living conditions and our fears for his safety.

Rosemary and I attended to our children as soon as they awoke and started making noises. We both feel it is important to talk *to* children rather than at or even down to them. My guidelines about how I would treat a young child are simply derived from how I would want to be treated by other people. Young children need to feel secure and protected to let their mind wander freely and explore their surroundings. Under pressure, or with the feeling that somebody has power and authority over them, children will not be able to develop their thinking and their capability to open their minds to new things and ideas.

Our children did things we didn't want them to do but we always tried not to get angry with them and never frowned at them. Toddlers are capable of understanding the various emotions without being yelled at long before they themselves can talk. There is nothing wrong with wanting to see what is in the kitchen cupboards, even if that means that the exploration of the cupboards requires the full attention of at least one parent to avoid breakages.

During these early years we tried to create a relaxed and loving environment with no aggression or harsh words. We didn't go out purposefully looking for things to stimulate our children but nor did we restrict their interests. While our children were all quite curious and very self-motivated, others might need more stimulation or deliberate introduction to activities. I would encourage anybody to do whatever their own interests and time commitments allowed. I don't think a parent ever runs the risk of overstimulating a child, as long as there is no compulsion or great expectations.

Although I am a strong advocate of starting children in maths as early as possible and preferably before or simultaneously with reading and writing, I would encourage parents to choose activities from their own range of interests during the very early childhood years. The first years are important to sustain the children's natural motivation and to supply sources of information, entertainment and excitement for their learning. But as I pointed out earlier, the activities don't need to be games. Rosemary and Michael had lots of fun hanging out the washing and counting the pegs, comparing the colour, size and shape of clothes on the line, or weighing the ingredients for a cake.

Admittedly, we had the big advantage that both of us worked from home since shortly after Michael's birth. This certainly gave us the freedom to respond to our children whenever they wanted our attention: to include them into our activities and to take time out for playing. While it doesn't mean that we spent every minute of our day with the children, we were approachable and available. Although parents working away from home might not have the same amount of time for their children, they may have the advantage of the energy of anticipation: higher-quality time together because they have all been looking forward to it.

Because of the large age differences between our children we could also give them more individual attention in the first years of their lives. I have often heard other people say that children need to learn their limits and boundaries and shouldn't be spoiled by their parents' constant availability and immediate, sometimes even unsolicited, attention. But a person's basic feeling of security can only come from the family or caregivers during their early childhood.

Child psychologist Glen Stenhouse wrote in a recent book, *Confident Children,* that a supportive environment and experience of a loving relationship are essential parts of growing up. These experiences are possible in almost any family, be it a single-parent family, a family living in extreme poverty or even parents living in divorce. I once taught a young boy whose parents were going through a separation process but had made an effort to explain their situation to their child and to create as much stability for him as possible. They took time to spend with him together and made him feel assured that both would still be available to him. It is this basic trust that fosters a lasting confidence and sense of adventure and exploration which are the most important prerequisites for learning.

Before children enter formal education, most will have

spent a lot of their time learning informally. Even in families where the parents had no intention to teach their children and made no deliberate attempts, children will have learned a lot. But when it comes to learning maths, many children don't get any opportunities because their parents don't quite know how to teach it.

Whenever parents bring their children to me for remedial maths saying they never felt competent to help their children themselves, I remind them of how much they have *already* taught them, mostly without even realising it. Automatically, without hesitation, they taught their children to speak, the biggest task a child ever has to achieve. Children start from scratch and within their first years they manage to comprehend the principles of language, learn to distinguish between words, start structuring short sentences, recognise the difference between referring to themselves or somebody else and gradually replace other, more basic communication methods with fully developed language.

Consider what you've already taught your child!

✔ to make sounds
✔ to distinguish between sounds, including grammar, vocabulary and how to organise sentences
✔ to paint and draw
✔ to sing
✔ to make up stories
✔ to use a knife and fork
✔ to help you in the garden
✔ to help with household chores
✔ to share . . .

. . . This list could go on and on; but maths, for all its importance, is rarely one of the things parents teach their children. Why not add maths too?

Maths books can be great story books

ANY FUTURE LEARNING WILL BUILD ON A CHILD'S language skills because we all need to express our thoughts in words, no matter if what we are thinking is to do with language or with maths. While most parents are happy to help their children to learn how to speak properly, and most even continue their offspring's expansions into the world of words by reading to them regularly, many seem to hope that school is early enough to learn numeracy.

I am convinced that these decisions are based on the parents' own preferences rather than some finding that children cannot relate to maths at an early age. On the contrary, recent research indicates that very young children are quite capable of understanding mathematical questions and using many of the symbols applied in maths.

Any teaching, be it maths, reading, arts or sports, must be geared to the child's age but all children have a hunger for learning which is not selective for any subject. Our own children were just as keen to do maths as they were to read a story with either of us or to have a game of badminton.

As I mentioned earlier, mathematics is hidden in many daily activities and tasks and often all you need to do is watch out for it and emphase it for the child. If you set the table, you could count the knives or glasses aloud so the child can hear a few numbers in the right order.

When you go shopping, why not give your child half the shopping list (with little drawings for the items you want to buy if they can't read yet) and talk about prices, amounts, volumes and weight.

A very popular and effective method of remembering the order of numbers is drawing dot-to-dot pictures. These exercises keep the child curious to see the resulting picture, they practise drawing skills and give children far more confidence in their drawing than simple copying of images would do. Furthermore, the child learns to read numbers in sequence. Dot-to-dot pictures are fun at any age and can be made easy enough for very young children, or be complicated enough to intrigue older children.

A beginner's example could be something like this:

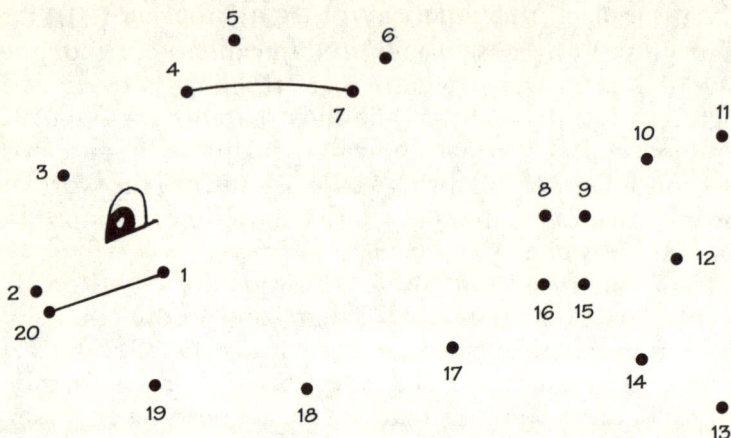

For older children who already developed an antipathy against maths it can also be helpful to combine the maths with something else they enjoy – a story, perhaps. Here is a beginner's version for children to practise their skills in recognising and distinguishing between numbers and letters. Take your child's favourite story to start with and, if you like the exercise, you will find that any story can be translated into a mathematical puzzle using basic mathematical operations or just numbers which stand for letters.

In a simple example, your child has to replace numbers with letters applying clues you provide to get the rest of a story. Find or make up a story interesting enough to make children want to hear it right through to the end. Here's an example:

Max and his friend Maxine were packing for a trip to the local zoo. The zoo allows people to stay overnight and to camp close to where the zebras and giraffes live and both Max and Maxine were very excited and looking forward to their night safari. They planned to meet four of their school friends at the campsite at six o'clock.

To hear the rest of the story, you can help your child replace the numbers with letters using the following clues:

1 = a, 2 = e, 3 = o, 4 = p, 5 = r, 6 = n, 7 = t, 8 = u, 9 = l (leave h, c, k, d, i, s, g, b, w, m, z, y, v, x, f)

Now replace each number with the appropriate letter and you will be able to read the rest of the story.

7h2y 41ck2d 7h2i5 7267, s9224i6g b1gs, bi99y 16d bi63c8915s 16d h3442d 36 7h2 b8s which w389d 71k2 7h2m 73 7h2 z33. 7h2y 155iv2d 726 mi6872s 73 six 16d d2cid2d 73 s7157 2277i6g 84 c1m4 wi7h387 wli7i6g f35 7h2i5 f526ds 73155iv2. 17 74m, 7h2y h1d 7h2i5 7267 16d 2v25y7hi6g 521dy 16d w252 w17chi6g 7h2 16im19s g51ziu6g 36 7h2 37h25 sid2 3f7h2 41dd3ck.

This kind of exercise has no limits and examples can be created from stories out of your children's books or some you make up together with them. The level of mathematical complexity can also be increased by replacing numbers with letters, to having to do calculations to get the right letter for a number. In this case, the clues in the above example could be: a = 5 – 4, e = 4 – 2, o = 7 - 4, and so on.

When should maths start?

ALTHOUGH I WOULD NEVER DENY THE IMPORTANCE of competence in literacy skills, I would also argue that numeracy is equally important, easier to teach successfully at an early age and a better basis for later acceleration or extension. Formal lessons at school often produce negative results and attitudes towards maths while an informal early introduction to maths will help children to experience the fun of it and will let them discover the maths hidden in everyday life.

From very early on we incorporated maths in the range of things we discussed with our children. On a walk we would show them flowers and count them or their petals. Even a simple comparison between flowers with either just a few or many petals will give a child an idea about quantity and comparison – both important concepts for mathematical operations. We started with more structured sessions at about the age of four, when they began discovering pens and paper and were quite keen to draw lines and shapes.

I will explain the steps of teaching maths to young children in one of the later chapters, but at this point I would like to underline that when I speak of early childhood education I am not referring to strict and disciplined early teaching like in the cases of John Stuart Mill and Norbert Wiener, or more recently Ruth Lawrence. All three were taught by their determined and single-minded fathers, following a rather strict regime of lectures and practice.

Mill and Wiener grew up last century and became very exceptional as children and subsequently famous as a result of their intellectual achievements. Both later published autobiographies delivering detailed accounts of their childhood years and pointing out some potential disadvantages of intensive early education. Although their intellectual skills and academic advantages were long-lasting and both men were regarded as exceptional scholars, they described severe difficulties around the time adulthood began. Both men suffered from a diffuse mixture of despair, self-pity, social ineptness and frustration. Mill claimed his deeply depressive tendency was caused by an education that almost entirely omitted feelings and emotions in order to sustain a focus on analytical thinking.

Ruth Lawrence is an example of a child prodigy of this century. She set a stereotype when she took a first at Oxford, aged thirteen. But wherever she went, her protective father would speak for her and make decisions on her behalf. Isolated from other children and exposed to a daily regimen of exercises and practise, she appeared to sacrifice much for the sake of her successful academic career.

This is not my way of helping children. At our home, the children initiate the activities and it is we, the teachers, who are happy to oblige. Michael was probably the luckiest as far as free choice of activities is concerned. He could freely decide whether or not he wanted to go to school, a luxury his older siblings and most other students don't have.

Michael is one of those children who seem to need a bit of background activity. He often worked on his maths in front of the television and had frequent breaks during which he would pick up his miniature basketball and dribble it through the living room and shoot at a small net outside. We were encouraging rather than pressuring him and, as he has never attended school before he passed his university entrance maths exams, we tried to offer him a range of activities including sports clubs and music tutoring, to give him a chance to meet peers.

Even with a self-motivated child like Michael, who has always been keen to have a learning session, it is vital for parents to know when to stop. Preferably the session should be over before it crosses the child's mind that it might be time to do something else. This can sometimes be difficult, especially during an efficient session when it is very tempting for a parent to think that the child could squeeze in even more information.

Mathematics has a definite advantage here because it is impossible, or at least very counterproductive, to carry on as soon as the child's attention is with something else. Partly for that reason, I never lecture my students plain theory. Their sessions are spent entirely on practising increasingly difficult examples. It is much easier to hold a child's attention when they actually have something to do, the sense of achievement is heightened enormously if they can give themselves a tick for a correctly solved exercise, and the learning and memorising of principles is more efficient during a practical session.

Although learning maths by pure practice can be highly repetitive, this is a very effective way of memorising basic operations such as multiplication tables which will form a solid basis for later performance in maths. Provided children have someone to help them with the exercises they can't solve easily, they will get into a sort of mental state which will take them through an impressive number of exercises, often without their even noticing that they are actually doing hard work.

During sessions I avoid looking over children's shoulders

– no matter how old they are. But you should never be far away if they need help or an explanation. As soon as you begin incorporating a bit of maths into your children's learning, it will fit into the day just like any other activity and there will be no need for any pressure. As long as children enjoy an activity they will come back to it. Depending on their individual interests and preferences, they will develop a passion for some activities and voluntarily spend more than the average amount of time on them.

For some children this can be extreme, especially when they have experienced a long period of failure and then just rediscovered the fun in maths. As a tutor for those who struggle with maths, one of the biggest compliments is when I get a call from a 'distraught' parent who tells me how difficult it is to get their children to close their maths books and come to the dinner table! But this is a frequent consequence of the rediscovered motivation and joy in learning after all the child's hangups about maths and previous failures have been cleared.

As with any other learning, it is more important to come back to maths regularly than to spend long hours on it at a time. Even a few minutes every second day are probably more helpful than long sessions once a month.

Personally, I wish children never had to go through demotivating experiences and could keep and even develop their natural curiosity without being held back or frustrated. Although young children might develop different preferences, I believe they all have an open and curious mind which is fairly indiscriminative at the beginning.

Anything offered to a child without pressure will be taken in – a process some argue could begin even before birth but certainly is quite obvious from the moment babies are born. All children can comprehend and do more at earlier stages if they have the right environment to develop their skills. Development of young children's basic skills such as language, reading and numeracy can be accelerated very substantially. The children who gain these important skills much earlier are likely to be advantaged in later life.

One example which surprised even an advocate of early learning like me, was a study into acceleration of physical fitness and ability. In this study two twin brothers were both trained in certain physical skills such as crawling up and down stairs, jumping and swimming. The only difference was that one twin was trained a few months earlier. He started at the age of seven months, while his brother received training only after he was nearly two years old.

The brother who was trained earlier was not only well ahead in all the skills practised but also progressed much faster through new tasks and skills. He swam on his own at ten months, could dive from the side of the swimming pool at the age of fifteen months and moved on roller skates and climbed steep slopes.

Environment and support are important

✔ Create a learner-friendly environment.

✔ Encourage the child to get involved with an activity, but be supportive, not pressuring

✔ Allow plenty of time and provide equipment like pencils, crayons, paper, rulers, etc.

✔ Start with more formal maths exercises as soon as the child can hold a pen and enjoys drawing lines and shapes.

✔ Remember that young children are not selective in their learning; whether or not they like a subject often depends on the presentation.

✔ Know when to stop – don't wait for the child to get bored.

✔ Don't hold lengthy lectures.

✔ Practice teaches much more than theory – do it *with* them.

✔ Encourage 'mental chanting', especially with exercises which require a lot of learning by heart (e.g. multiplication tables).

✔ Don't worry about 'burnout' as long as you don't overdo it. Not to offer your child the opportunity to learn is to risk 'rustout'.

I was surprised to see that his advantage lasted well into adulthood and he always displayed better co-ordination and more confidence at physical tasks. I couldn't help thinking that if there was a chance to give a child an advantage there should be nothing to stop parents from doing so. The most positive outcome of this particular study, however, was that both brothers had developed above-average physical skills with the differences between them being much smaller than the general advantage they had compared with other young men.

There are many examples like this, and the knowledge that children can reach much higher skill and at the same time have fun always encouraged me to offer what I could to my own children and to my students. People have often criticised me saying I was risking my children's 'burnout', and I agree that there is a risk of their losing motivation and direction as young adults because they have achieved all there is to be achieved and may run out of new challenges. I am much more distressed, however, by the thought that we risk our children losing motivation much earlier by 'rusting' them out. It takes much more energy and time to help a child to regain their enthusiasm than it takes to sustain their inborn drive and urge to discover and understand.

The parents' or caregivers' role

PARENTS (OR CAREGIVERS IN THE CASE WHERE A child is not being brought up by its parents in the strict sense of the term) should never be too demanding. The motivation always has to be to share knowledge and information with the child and to open avenues for their explorations, but never to mould a young, open-minded person following some fixed idea about goals and achievements. I have met many parents who were convinced they 'knew best' and who wanted their child to be and achieve what they haven't managed to do themselves. This may not be uncommon, but we should always keep in mind that each child, is always a new person with differences and special qualities.

We are always happy if our children decide to follow other interests. In fact, all our children had sometimes prolonged periods when it wasn't clear whether they would ever take up mathematics again. Audrey was so enthusiastic about her music and dance that she began contemplating a career as a ballet dancer. We were also proud to see David being selected for the Canterbury Under-12 soccer team when he was eleven, and we would never think of stopping Michael from being keen on tennis and spending hours playing the piano, or even watching TV.

All these other passions are necessary and they provide experiences that are just as valuable for the child's development as anything they learn and experience at home or at school. The more children excel at one particular subject the more they need the balance of other activities – not only for their own growth but often also for their image outside their home. Many people, including children, can be intolerant or even hostile towards those whom they perceive as 'odd' and peculiar. There is a definite bias in society towards certain abilities and achievements. While a teenage sports achiever might have a nation of admirers behind him or her, somebody who excels only in the classroom is often much less enthusiastically received.

I would always support each individual's choices even if some children develop their strongest interests in purely academic fields. I can't see why there should be a different value in achieving in different areas provided the child concerned is enjoying it – the individual's level of satisfaction should be the yardstick. As long as learning continues to be regarded as a competitive task, there will always be an element of isolation and distance which academic achievers will have to cope with.

Although there is a certain level of competitive thinking and behaviour in all of us, children are not naturally competitive. Achievement does not have to go hand-in-hand with a competitive environment: on the contrary, children will develop their skills easier if they are supported on a more individual basis.

Everybody who lives in a family with more than one child

will know how damaging comparison between the siblings can be. While this situation is probably more damaging because of the closeness between siblings and the emotional needs of acceptance and belonging, competition between children at school is simply a less harmful version of the same principle and will only serve to motivate those who happen to be considered better at a specific task.

Many essential social skills also can't be learned in a purely competitive environment. Child psychologist Glen Stenhouse writes in his book *Confident Children* that although regular assessment of progress is an important part of school education, an inevitable result is the comparisons students make among themselves. Young people receive frequent messages about whether they are succeeding and how their performance compares with their peers. He says students are likely to take their marks as an evaluation not just of their performance in an isolated area of skill and knowledge, but of their total functioning as a person. A child's overall self-concept will be affected accordingly.

For all these reasons we try to prepare our children to rise above potential conflict, jealousy and aggression and we always make an effort to incorporate more emotional qualities in the home tutoring sessions.

Humility and helpfulness would be among the most important qualities we try to instil in all children who are taught maths by us. Children who considered themselves as failures for a few years, then unexpectedly find themselves a success, can be particularly susceptible to boasting, arrogance and show-off behaviour. To a certain extent this is a natural way of showing pride and self-satisfaction. But a much better way of achieving this state is by getting involved with teaching and helping others.

Most children quickly realise that they can be happy about their own achievements and popular with other children at the same time if they learn to share their knowledge and help without putting the recipient down.

Even with groups of average classroom size I would never single out a child for either impressive accomplishment or failure. I would never let a child struggle trying to work out

Basics for progress

✔ Don't leave your child frustrated.

✔ Don't leave them feeling incapable of learning.

✔ Give as much help as you can and as your child is happy to accept.

✔ Children need to be confident of your love with no strings attached – your child's performance should have nothing to do with your love.

✔ Don't compare your child's performance with that of others.

✔ Always accept the current level of performance as personal best at the time.

✔ Encourage children who make quick progress to help others, not to be smug and superior.

something for themselves, although many educationists say that it might be the better way to learn. I always encourage learners to ask for help rather than leave them frustrated.

My experience with learners has taught me that self-motivation develops alongside support and understanding, and will gradually take over if a child feels secure in the knowledge that somebody will help if necessary. They soon begin working on their own progress from one personal best to the next, moving from strength to strength.

Looking back, I often think Rosemary and I grew into an efficient team because we combine qualities which make it easy for children to learn. Rosemary is perhaps an unusually committed mother. She has a talent to make everybody feel at ease and welcome, is always accessible and has a strong sense of humour. I seem to be able to pass on my enthusiasm for learning and for maths in particular, and I have learned to be patient enough to give each child enough time to proceed at their own pace.

Can anybody do maths?

Our students arrive with a wide range of problems and casualties of both private and state schools have ended up at our doorstep. For various reasons, they have either never enjoyed maths or stopped enjoying it after a sequence of failures. But most of our students' so-called failures arise because they have tried to fulfil their teacher's or parents' expectations and became increasingly stressed, frustrated and anxious. Many have been pressured into taking up private tutoring and arrive in their first session with low motivation.

Sometimes all it takes to get them started is to tell them that it won't take long and won't be difficult to fill the gaps in their knowledge, and that nobody expects them to know more than they already do. Even a child full of anxiety will accept that there is no real problem and nothing to worry about seriously. It might not be enough to motivate a child to do maths exercises but it sparks either relief or the urge to prove me wrong. Either way, the child is curious to find out what comes next. The prospect of working through material taken from textbooks for older children can also be quite motivating for some, but should never be applied if it puts more pressure on the child.

When I tell them that they will never be left unable to do their maths again, most children are puzzled but again they are keen to see how that should work. These basic principles rekindle whatever general motivation is left in a child, and to a varying degree this works for all children. Having taught children from extremely varied backgrounds for twenty years now, I am convinced there is no big difference between their potential performance and level of achievement, no matter what their background.

Maths is surrounded by many myths like the belief that girls are not as good as boys, poor people are not as clever, and so on. I don't believe that there are *innate* differences in ability and capability, but rather *enforced* differences based on preconceived conceptions about the potential level of achievement. Survey statistics might show lower participation in maths by girls and a ranking of participation and performance by members of the various racial backgrounds. However, I believe these figures only serve to reinforce the stereotypes which are responsible for these differences, and have no value when it comes to predicting a child's potential in a supportive environment.

With regard to their academic achievements, there was no noticeable difference between our sons and daughter, and even after several hundred students came through our home I could not detect any significant gender difference. Individual differences were always much greater.

I am pleased to see traditional gender roles changing but, as with any major change, the younger generation seems to adjust a lot faster. It can still be more difficult for a girl to receive the same support to venture into maths and science from a teacher who grew up with more conservative values. At our home, we never thought about treating our daughter differently and I cannot imagine why the potential to learn should somehow be different in boys or girls. Individual differences in temperament and personality, however, certainly affect how easily and happily a child takes to any particular subject.

Research on identical twins who were separated after birth shows that some qualities contributing to the personality of a child are probably inherited and ultimately may have an influence on the child's development. Pushing a child into a field for which it is not temperamentally suited is a recipe for disaster but, on a more general basis, any child can be supported when it comes to sparking their curiosity and increasing their attention span, their perseverance and patience, or their assertiveness. All these qualities are essential for a young learner, and teaching maths can support their development more than most other subjects. Although these qualities might have a genetic basis, none of them will be developed in a rigid way and they can be modified significantly during childhood.

Racial prejudice seems to be spread more widely and often sits deeper than gender prejudice. Being of Malaysian Chinese descent, we often came across this problem ourselves. There seems to be a persistent image of Asian students as shy, bespectacled, with a high level of discipline and very hard working. A good education is highly valued in Asian countries and it is becoming increasingly difficult to place children in schools and universities. I would agree that there are cultural differences between children from various backgrounds but, as before with gender specifics, I find greater differences between individuals than between members of any racial group.

For Maori students there are other stereotypes ready to be pulled out should somebody dare to be different, but all these descriptions are worthless and potentially dangerous, and teachers should always make the effort to get to know the individual.

I taught a group of Maori children of Ngai Tahu descent for two terms and experienced first-hand how common these negative images can be. One of the girls I taught improved from scoring 42 percent beforehand to 90 percent after she finished the course but at school she got accusations of cheating, as though a Maori girl could not possibly improve by that much.

After two terms of teaching the group, more than 70 percent of the children had made significant progress and most showed across-the-board improvements in their marks in many disciplines, not only in maths. However, I was still disappointed because I would have liked to see them all pass School Certificate, but many had come from so far behind that this would have been unrealistic. Their home and social behaviour also improved, to the delight of everybody around them.

Perhaps the best result of this pilot project was that the children actually started coming back to school voluntarily. Truancy had been a problem before, especially with the older students, but we chose to use a university seminar room as the venue for the sessions and the setting proved a very powerful source of inspiration for the young people. They had been given a new perspective and an environment where nobody doubted their ability.

We gave them all the time and help they needed to get over hurdles, and even the most notorious truants were

> ## Everybody can learn and enjoy maths
> ✔ in a non-competitive environment
> ✔ at their own pace
> ✔ when they're in control of their own progress
> ✔ if allowed their own choices of topics
> ✔ with encouragement and help readily available
> ✔ irrespective of race and gender.

suddenly returning to the classroom. The students' general feed-back was that they felt much more in control of their own learning because they were allowed to make their own choices and move through the material at their own pace. They said the lack of any formal tests or assessments made them realise they were learning for themselves rather than for the teacher.

Parents also told me that their children's self-esteem improved markedly. While they didn't seem to be interested in participating in any school activities before we started the sessions, they became very interested in tertiary education options and were much less affected by negative comments about their academic potential afterwards. They had learned to believe that they could make a change to their lives, disregarding what other people thought.

The group had chosen maths as the subject because it had been a concern for Maori students, who were consistently lower achievers and generally did not appear to show much interest. Yet overall, maths was just a vehicle to help them regain their curiosity and confidence. The children's improvements in maths were secondary to the fact that they changed from feeling miserable at school to looking forward to the next session and knowing that they could in fact achieve beyond their own expectations.

The programme director decided to extend the pilot scheme, and began training parents and teachers to provide ongoing maths tuition.

This experience confirmed my belief that all children and young people are able and keen to achieve far beyond their parents' and even their own expectations if they are taught in an non-competitive and encouraging environment

The brilliant and the ordinary

BRILLIANCE AND EXCELLENCE CAN TAKE ON MANY forms, and most people would know somebody they consider to be brilliant, gifted, talented, or simply really good at what they are doing. Unfortunately, most people think they could never match that level of achievement and take refuge in the comforting idea that only a few can be so successful.

High achievers seem to have some special advantage which propels them far beyond anybody else's level. On top of all that, they often even seem bold enough to enjoy what they are good at – they insist on doing only what they enjoy, which is something many people feel uncomfortable with. Depending on what other social and communication skills achievers bring with them, our response to them can range from respect to resentment, tinted with a little jealousy.

I don't believe that anybody who happens to develop better than average skills in any task is in any way advantaged from the start, but some people just seem to have better opportunities to make the best use of what they bring with them. To me, innate, genetic factors are of little significance for developing abilities to the highest possible level. Innate factors might well influence our temperament or personality but never in a rigid, inflexible way. Development and change can always be achieved by choice and the decision to make use of an opportunity.

Many people described as geniuses have told us that the mind is capable of achieving much more if a person is provided with the freedom to focus and concentrate on a particular task and the opportunity, encouragement and support to give one's full energy to it.

Young children do exactly this. They take the freedom to concentrate fully on what ever they are doing at any particular moment for as long as it takes for something else to grab their attention. Their desire to discover and learn new things is so great that many parents have set out deliberately to create a 'genius'. But one must always keep in mind that there is a very fine line between helping children and desperately wanting them to achieve and to live in their reflected glory. Pushy and pressuring parents can have a very destructive and unhealthy influence.

Parents or caregivers will always have the most influence on a growing child but I believe children themselves should be able to dictate their hunger for learning and the parents' task should be to support them along the way and try not to hold them back. Parents know their child better than anybody else and the basic trust children and parents have for each other is the best basis for a supportive relationship, free of pressure and obligations.

But even if we all decided to support our children and provide more opportunities for the next generation, parents should not have any preconceived ideas about what the child should be able to do. Even if all children receive more encouragement and freedom to learn, there will still be differences in the pace of their progress and the activities they prefer. I think parents should work with their children, keep an open mind and take whatever time it takes to discover their child's gift, rather than determine that the child should be accelerated and work towards a specific goal.

Until recently, excellence was measured through tests based mainly on what we describe as intelligence. Despite the fact that most people could probably describe intelligence only with very vague terms, most of the assessment of children's capabilities was based on IQ tests. Later definitions of giftedness include areas such as creativity and task commitment or perseverance and motivation, but still assume an innate difference in children's capability. Most of what we consider as gifts and talents, however, is based on our current attitude towards the particular task achieved rather than any special abilities of a child. If we place high value on academic achievement, or regard it as very desirable when young to be talented in sports, music or any other area, then we will think that somebody who can do that is 'gifted'. Definitions and methods of identifying those who are gifted will continue to change and I will always feel uncomfortable with the idea of separating children into groups reflecting their level of performance. I would prefer to give as much support as possible to each individual while *disregarding* their special interests and qualities.

What we measure as intelligence has not much to do with the potential of a child to enjoy and achieve in later life. Many people regarded as highly intelligent when they were young still end up in destitution as adults. On the other hand, there are all the anecdotes about Albert Einstein failing miserably at school. Intelligence and later achievement are far from perfectly correlated: more essential qualities for achievement are self-confidence and freedom from feelings of inferiority, as well as persistence or 'stickability'. Together with a generally high motivation and drive to discover new things, these qualities are even more important than prior knowledge or experience.

Mathematics, at least in the school text books I teach from, always has the aspect of solving a puzzle knowing that there definitely is an answer. The sense of achievement and satisfaction is very high in mathematics because successfully solving a problem is somehow immaculate and pure. Nobody can unfairly criticise a correct answer!

If children were introduced to maths in a supportive environment their experience of the subject would be only positive. Each of their tasks would have a definite solution and thus always a final goal. Although there might be several ways of approaching the problem, each step they took would get them either closer to the result or provide further hints as to the next move. A supportive parent or teacher would discuss the chosen pathway of solution, rather than just determine whether the step was right or wrong. Children would learn a higher degree of self-determination and could perform real choices – all of which is very motivating for a child.

Intellectual activity can be especially rewarding if it is guided by somebody more experienced, and as long as the mentor or teacher avoids putting the children down it will develop their independence. Maths is highly objective and allows for almost no argument over the results. Many of our negative school memories go back to English essays or analyses of some question in history which we submitted proudly only for the teachers to mark according to their own prejudices. Mathematics is a 'safer' subject in the sense that it gives a child the opportunity to receive unconditional praise.

Even parents who feel unable to help their children are probably best qualified to support them in developing self-determination and independence. However, excellence in any task should never be pursued just for the sake of achieving beyond the average, but must be seen as an enrichment of the child's future life. It therefore also has to go hand in hand with experiencing and learning other skills that foster acceptance within the child's peer group. For us, humility, modesty and helpfulness have always been very important qualities which we wanted to instil in our children from early on. We wanted them to use their skills to help others rather than feel superior.

Although recent findings by educationalists indicate that all children can be accelerated and can acquire certain valuable and basic skills much earlier than most normally do, still many parents and teachers often hesitate because they are not sure whether the advantages outweigh possible unforeseen disadvantages. What is the point of accelerating children if they are likely to encounter hostility and envy, or perhaps suffer early burnout and isolate themselves from their natural environment and friends?

To me, all these fears are only relevant if support and acceleration was only available to a small group of children who would not be allowed to pass on any of their knowledge and achievements. The prime emphasis for us was always to give each child a chance of developing as far as they can go towards a happier learning. I often feel it is still ideologically unfashionable to defend the notion that children can be given a happy head start, but the evidence we have seen through our own and other children shows that most young achievers live fulfilling lives and are happier than those who constantly fear and experience failure. Early achievement makes children more independent, purposeful and positive and just as well socialised as others.

Skills developed early should not be overly attributed to a child's innate mental activities or abilities. I often come across the misconception that children who read earlier, faster and easier or comprehend maths earlier and better are somehow 'brighter' than others. This belief is often held all the more strongly because nobody in the family made any conscious effort to teach the child, and it seems that all the special skills simply developed naturally. But perhaps the early readers had the advantage that their parents, often unwittingly, taught them to distinguish between sounds, taught them about words and the written language by talking about it and thus created an atmosphere of expectancy and curiosity. They may have given their children something to look forward to when they finally learned to read books. Without realising it, the parents gave their children a head start when they first approached more formal reading lessons.

The same advantages can be achieved in maths, and any other topic or subject, with similarly striking results, which often misleads people into believing there is some sort of special talent at work. In maths, the concepts of quantity, size, space and shape are easy to introduce early and will give a child an advantage when it enters the school system. If the introduction to a subject happens informally and unintentionally, many parents are not aware of the advantage they have given their child and are later surprised to see more confident and competent performance at related tasks.

I would say it is within the means of any parent with time, care and patience to give their children the kind of early start in life that will vastly increase their child's chances of becoming a well balanced, able and happy young person who will also have learned to deal with negative experiences and hostile reactions from others, without being arrogant. It is important, however, to prepare them for these negative reactions in a way which helps them to understand the origin of envy, jealousy and rejection.

We always encouraged our children to help others and to pass on their knowledge or understanding. I could never see much sense in simply accumulating knowledge or learning new concepts just for the sake of being better or faster than others. Unfortunately, this competitive aspect of learning is still widespread, and many parents' ideas about accelerating their children derive from the ambition to better others.

Although our children achieved certain goals and levels of knowledge earlier than others, they were always aware of the fact that anybody was capable of doing the same and they were happy to help anybody who struggled. David was more popular than I was with many of our students because he was so good at making children feel they could catch up with him. And many did so without wasting any time doubting their ability. Because of my own focus on maths, our children also needed other teachers to coach them in all the other subjects needed for school certificates and university entrance, and they learned that their own knowledge could always grow and be extended. They also discovered how important it was to share knowledge and to keep an open mind about any subject or idea.

In the classroom, it should be just as important to support children's own confidence in their abilities and skills. But with groups the size of today's average class it is undeniably difficult to give the same attention and support to each child, and many will miss out on the early encouragement they need.

Most schools also work on the principle that there is a common goal to be achieved by as many children as possible each year. While support programmes exist for those who struggle, far fewer schools offer support for those who could do more or are keen to investigate subjects in more depth. Personally, I think we should make an effort in offering individual support to all children and allow them to move through material at their own pace – even to a certain extent selecting their own material. Children who work on a subject they are really interested in, even if this means they spend some time working outside the curriculum, have the best chances of developing good basic learning skills which will make all future learning easier.

Yet all these issues come up long before a child enters

school and parents have to decide how much they want to support their children's development and how they can balance their influence with each child's temperament and preferences.

A study into a group of extremely competent young musicians found that none of the children showed any early signs of special or unusual ability. In most cases it was parental support that made all the difference. The results of this study did not surprise me. Our own children also never really stood out as youngsters until they developed a passion for their subject and decided to give it a lot of their energy. We introduced them to maths and many other activities and they determined how much time and energy they would devote to each of their interests. Our encouragement was always there, whether they asked for more and new information, or for someone to play a game of table tennis. Neither innate talents nor parents with high expectations will enable a child to achieve what a supportive home environment can do.

But even if we provide optimal early childhood opportunities and conditions, children will differ in their interest, self-motivation and general attitude towards learning and achievement. And of course, in almost any society there will be some sort of ranking order for the various abilities and specialities, and derived from that are people's attitudes towards a particular achievement.

According to researchers who looked into the historic context of attitudes towards high achievers in different task areas, this 'prestige hierarchy' has remained relatively stable for centuries. The highest level of recognition seems to be reserved for those who combine several rare skills and take on leadership roles. Creative scientists who consider the social and economic impact of their findings; political leaders who manage to incorporate the many facets of society into their thinking; determined and enthusiastic campaigners for a widely recognised cause and those who are able to filter vast amounts of information into something meaningful for the ordinary person, are ranked high on the scale of ability.

Compared with this, skills and interests in the creative

and artistic fields are considered more as a surplus by many people. Writers, musicians, painters and many other artists focus on skills that are not considered essential for the individual or a society to function and progress, but are still valued for the pleasure they can bring to those who develop them as well as those who enjoy the artists' products.

Less recognised or appreciated are skills at the extreme ends of the spectrum. Manual skills and the ability to maintain concentration at mundane tasks on the one hand, and very high or abstract fields like academic research on the other hand tend to be downplayed by people who see themselves in the 'middle ground'.

Many more skills and talents are often unappreciated or disdained. Performing calculations faster than a computer, speed reading and memory tasks are often regarded as gimmicks rather than special abilities and can earn the performer a range of responses.

Academic achievement is generally highly regarded but often met with a lot of suspicion when the high achiever happens to be a teenager. It smacks of elitism and as long as learning continues to be based on competitive principles, there will also be a difficulty in providing children with the necessary support to develop their skills.

Children who are lucky enough to live in a supportive and loving family will learn about this competitiveness as soon as they start school. Although things are changing in schools, there is still a lot of comparison between individual children and pressure to work towards a common goal. Whoever gets there faster or more easily is regarded as more intelligent. If we are serious about closing the gap between the successes and failures, we need to look first of all at the competitive concept of learning. This comes with the implication that achievement of a certain task will earn more recognition and acknowledgment than in other areas which might be more attractive for a particular child and come more naturally.

Although schools and education departments are trying their best to incorporate as many opportunities as possible, many children are still having a miserable existence at school and, despite their potential to develop interests and desire to achieve, they become increasingly convinced to be a failure simply because their passions might lie outside the range of school subjects.

I have many students come to me for remedial maths because maths is the subject where their failure is most obvious and where they feel most intensely inadequate. It is a very common experience for me to start with students who have been through years of difficulty at school, experiencing failure in tests, rejection from classmates and the ridicule of teachers. All that just because they failed to develop an interest in the general school subjects, failing more obviously in maths than in other subjects because it is a less permissive subject, but mainly because they never received any encouragement for things they could do well.

Although mathematics is often the last thing they are genuinely interested in, it has also been the most frightening and often disabling subject and they feel unable to develop new skills or tackle new challenges without first resolving the problem in maths. So let us look at the mysterious task of teaching mathematics, a subject which seems to present some of our children with unconquerable difficulties.

What's the secret?

MANY PARENTS AND TEACHERS WILL BE CURIOUS to hear the details and secrets of my method, Rapid Mathematics. Yet, reading this chapter, some readers will be reminded of their own good intentions and commonsense approaches. If there really is any great 'secret' to my method, it is to find the energy to follow your good intentions consistently and never become impatient.

My approach to teaching has evolved over twenty years' work with children from pre-school to university entrance level and is still being further refined. It is not based on textbooks or scientific findings about early childhood development. I rather look at it as something that I began by pure chance and developed gradually over the years.

A few basic principles crystallised out, but even these should not be regarded as hard and fast rules – they are mere guidelines and will need further adjustment for each individual.

A good way to start teaching a child is to actually *believe* in their potential to progress, develop and acquire skills. This might sound like a rather obvious statement but I am sure everybody will remember some encounter with somebody who managed to undermine our confidence to do something just by not believing in us. Children often have the ability to sense this in adults.

You should work from the premise that all children can understand even complicated concepts if they have the time and encouragement to keep going. My belief in my children and students comes quite naturally simply because I have no reason to assume otherwise.

A balanced mix of encouragement to recognise and appreciate children's natural motivation, and regular doses of satisfaction applied through little successes along the way, may sound a simplistic recipe for fostering high achieving. But I believe these principles are basic for creating an environment in which children, and adult learners, will develop to their highest potential. And they are too easily neglected under mounting pressure.

How we started
David was seven years old when we started regular sessions after school. We first practised what he did at school at that time – basic sums and subtractions. Initially, I didn't think of extending him but he was quick to pick up new things and we simply moved on to successive chapters in his textbook. I trusted my intuition to tell me the right pace, to skip topics or exercises whenever I felt David understood the concept, and to encourage him enough to stay motivated.

Because I had a working knowledge of maths, I knew how to move rapidly through the book without boring David or demanding too much of him. We only did a few exercises, which never took longer than half an hour unless he really wanted to do more. I watched him very closely and, when ever his thoughts seemed to drift away, we stopped and did not continue until he asked for more.

At this age David had no particular preference for any subject and was quite keen to learn about anything. As he was my first student, it was very important for me that he should not lose the fun he had learning. During standard two and three I taught him most of the basic primary school work and in standard four, at the age of ten, we began with calculus. I then suggested he could attend night classes to prepare for the university entrance maths exams, which he passed the same year, 1977.

With Audrey we began a little earlier. Rosemary had sought story books to read to our children that introduced numbers in an imaginative way. Their favourite was *The Very Hungry Caterpillar*, but there are many others (see page 112). Audrey was about five years old when we began with maths, and at the age of seven she was doing fifth and sixth form maths. In her case I had moved through the exercises and topics a bit faster, gradually increasing the complexity of the mathematical questions rather than trying to cover the whole ground.

I worked on the principle that I had to make sure she understood a concept and that she could work on the exercises whenever she felt like it. She had a sound knowledge of the subject at eight and could have passed UE maths then. But she did not feel ready to sit the exam – a decision we left up to her.

With Audrey, I learned what enormous amounts of information very young children could take in provided they were enjoying it. Although Audrey's early training was rather specialised in maths, it still taught her many other things that helped her later with other interests. She followed her music and ballet with the same focused concentration she had learned during her early maths sessions. She had fun and returned home from ballet practices relaxed and satisfied that she had achieved very well.

Michael, our youngest son, was entirely home-taught, with a strong focus on maths. Because he was so young, we left it entirely up to him to do as much as he wanted – and *when* he wanted. We often had to fit the maths teaching in between his favourite TV programmes. After he passed his form seven Maths with Calculus exam at the age of seven, in 1991, he wanted to try school and divided his time between primary school and the university's maths department for a while before deciding to home-school when he turned eleven. On the day he passed his bursary maths exam he was far more excited about the cricket bat we gave him than about all the media attention he received!

To many readers my family's story will probably sound a bit like experimenting with children, which strikes a chilling note with many. Yet my own and many other children I taught experienced a major improvement of their self-esteem and confidence, and often also improved their social relationships, both at home and with their friends.

Teaching maths as a parent

IF YOU'VE NEVER IMAGINED YOURSELF TEACHING your child maths, you might be surprised how much you too will learn from it. Try to leave behind your own fears about maths: they will only hinder you in your new discovery and they are all too easily passed on to children.

Gaining further knowledge and experience is a very rewarding process and I always felt that knowledge is a privilege – even more so if you get an opportunity to share it. Your own enthusiasm about learning and understanding will soon be contagious.

Try to keep the focus of your teaching on the individual, however many children you are teaching. Parents usually know their child better than anybody else does, and will be the first to notice if they are bored or starting to lose concentration.

Most children go through a phase of acute curiosity during which they can drive their parents up the wall with their constant flow of questions. It is important to encourage them in their explorations even into subject areas in which we might have developed a dislike. It is more difficult to win back motivation once a subject has been spoiled for a child, but we all seem to retain at least a certain amount of curiosity which, in fact, is nothing less than motivation to learn.

The first thing you have to establish in the relationship with your children is that you are not just another teacher who will mark their tests but somebody who will help them clear their problems or extend their knowledge. Some children need time to adjust to that because it may be the first time they have had an opportunity to learn without being under pressure to perform or pass a test. Any form of assessment will make children feel under pressure, and the focus of their learning will shift from taking in as much as they want to know, to learning everything superficially for the short term to pass the examination.

Your focus should always be on encouraging children to get involved in their work and subject and to enjoy their progress, rather than learning the material to please you or pass some test.

It is essential that your child realises that you are not just trying to prepare them for the next test but really want them to feel good about their level of skill and achievement.

More important than the maths is:

✔ to believe in your child
✔ to remain patient
✔ to focus on each individual
✔ to keep each lesson short
✔ to have some rules but be flexible about them
✔ to encourage exploration
✔ to reward through ongoing support and unbroken encouragement
✔ to skip assessments and tests
✔ to remember the teaching is aimed at benefiting them, not at making them live up to your own expectations
✔ to regard it as a learning process for yourself too.

Ensure they understand clearly it is *they* you care about, not their mark in the test.

This teacher/pupil (or parent/child) relationship seems crucial to me, whether you want to teach your own children at home before or during their school years, or to extend them beyond their school work. No matter how young or old the child, they have to feel secure in the knowledge that there is someone who will help them with whatever their problem is. And this has to go beyond any particular subject.

Not just a teacher but a friendly coach
When I introduce myself to children I often use analogies borrowed from sports. My favourites are drawn from hurdling, and the Scottish game of curling. I am there to sweep the path in front of the stone so it keeps going, or I will push the hurdles away so you can keep running. The support has to be ongoing and unconditional – no matter how well or poorly students perform. In a family, this kind of relationship might come quite naturally, although some parents are convinced their child is 'special' and may have egocentric and competitive goals.

I use slightly different approaches when teaching preschoolers or extending school children. Although most of the children I teach are primary school children who come for a two-hour maths session each week, I would like to start where all teaching starts – with very young children.

Coaching pre-school children
How can all this be applied to early childhood training or home schooling? One would hope that many of the characteristics I used to describe my relationship with my students would be close to the natural relationship between a child and a supportive parent. But I realize many parents don't have the time, patience or confidence to teach or guide their children. Particularly in maths many feel totally inadequate. Unless they have dealt with maths professionally many feel their own knowledge is too fragmentary to assist children, even at early primary stages.

Forget these negative feelings and look at learning with

your children as a challenge for a second attempt at the subject.

Although most parents are very concerned about their child's literacy and read to them a lot, the vast majority of children are innumerate before school and some still lack basic numeracy skills when they leave the education system. As adults they are denied access to a lot of information just because they can't read a graph, interpret statistics or even check figures in contracts or bills. Even more importantly, their confidence and self-esteem may be lower than in people who have mathematical skills or who never built up negative attitudes towards learning.

All children are inquisitive by nature. They might have different preferences for methods of taking in information and might respond to different topics with more interest or less, but generally I don't think a young child can be bored. Everything is new and therefore stimulating, and learning happens almost automatically – without being under pressure to perform and without being tested and marked at the end.

The longer parents and caregivers manage to sustain this learning atmosphere the more the child will take in and retain; the more they will enjoy learning. However, in many families even the commitment to answer the child's questions as patiently and as promptly as possible is hard to uphold. Demotivation probably begins very early for most children: when they first hear: 'I don't have the time now.'

I don't always approve of early intensive education, especially if it is initiated by the parents rather than the child, but why shouldn't we include songs about numbers, counting games and the drawing of circles and triangles in the daily repertoire of activities?

Although I would incorporate maths in a child's early learning with at least the same prominence as reading and language skills, I believe the really important issue for parents is to do whatever will keep their children interested and motivated. When a friend who had no particular interest in maths himself asked me how he should best teach his three-year-old daughter to make sure she was well prepared, I told

Some ways of keeping your child curious about maths

Magic numbers: Use a candle to draw an invisible number and the corresponding number of objects on a piece of paper (e.g. a 3 and three animals.) Then let your child paint over the sheet of paper with a dark colour.

Jigsaw puzzles

Counting rhymes: For example:
One, two, three, four, five,
once I caught a fish alive.
Six, seven, eight, nine, ten,
then I threw him back again.
Why did you let him go?
Because he bit my finger so.
Which finger did he bite?
This little finger on the right.

Songs with mathematical ideas or sequences: *Ten Green Bottles* or *Green Grow the Rushes-oh*.

Musical bottles: Fill three bottles to a quarter, half and almost full with water and try to produce a tone with each. Dicuss the difference you hear in each tone.

Stories that have to be calculated to hear the end.

Rising raisins: Take a glass of fizzy water, drop in a dried raisin and watch. Gas bubbles will form on the raisin and lift it up. Count how often the raisin floats to the surface and sinks again; measure how high it rises, etc.

Anticipate an important day and count down the nights.

Count all the circles or triangles (or other objects) you can find in a picture.

Feelie boxes: discover different shapes and surfaces.

him to involve her in everything he did and only teach her whatever came easily.

Although adults seem to have preferred ways of taking in information – for example, acoustically by just listening to somebody's explanations, or visually by watching events – children need to be exposed to as many approaches as possible. Simple maths can be provided by coloured beads that can be counted into piles of the same colour; by counting songs; by counting the flowers or leaves on a plant; by looking at the various shapes of objects in the kitchen; by bringing in and counting the milk bottles. There are lots of ways to include simple examples of maths in everyday family situations without putting children under pressure; for example: 'I wonder how many milk bottles we need to buy if each of us drinks one a day?' could be a question to ask a young child on the way to the grocery store.

We always tried to give our children as much information as possible without lecturing them, while providing more detailed explanations if they ask. The children were always included into everything we did. We treated them as equal members of the household with every right to know what was happening and why.

One way of checking learners' interest and understanding that I employ daily with Michael is to answer his questions in a prompt and friendly manner. If he asks more, I know I have not bored him with over-expounding, but I proudly notice his confidence to tell me, 'I've got it, Dad,' if I start over-explaining.

More formal teaching sessions were always initiated by the children. Maths has the big advantage that it cannot be forced onto anybody. You have to feel like learning maths to be able to take in any of the information presented. So, my most important job was to spark the interest and to keep them motivated. Just as when you might announce tomorrow's story to keep the child curious you can prepare little teasers in maths as well.

Parents as first teachers

Parents will always be a child's first source of information and inspiration for everything including learning. Often unwittingly, parents introduce their children to things around them and spark their interest in things they themselves might take for granted. Children are constantly processing new information, learning to understand new concepts and drawing new links and connections between bits of information they come across.

Teaching and learning does not start on a new entrant's first day at school, but his or her first day of life. It is an ongoing process in which parents take the role of the first and most respected and trusted teacher. The easiest way to learn new things is when children can take their time for discovery. Instead of setting out to teach a very young child something specific, parents can probably help their children more if they encourage them to ask questions and support them when they begin to experiment with things, ideas and concepts. Parents also have to be prepared to offer some explanation and guidance through the maze of information.

While we all agree that children need to learn how to speak, to move, use things appropriately, learn basic behavioural skills and develop the characteristics of their personality very early, mathematical skills often seem too academic and generally less useful. I have often heard the argument that parents do not know how to start with mathematical skills: how to teach maths without making it a boring chore.

If we teach a child to speak, we almost automatically apply all the techniques that make a good teacher. We point to something, say its name, repeat the word as often as possible, let the child see, smell, feel and hear it and come back to the same item again and again. Imagine you are explaining a flower to a young child. You let the child smell it, touch it, caress the child's cheek with the flower to let it feel the velvet of its petals – and doing all this you talk to the child, repeating the words as often as possible. Each time you pass a flower you do the same again and let your child absorb the information. However, although you might expect your child to recognise the flower as something familiar, you would not expect them to be able to say any of the words immediately.

Introducing young children to mathematics is no different from that. You can count blue cars on long journeys, you can count pebbles or shells on the beach, you can sort leaves by their shapes – and all these activities can be repeated whenever opportunity arises, without expecting your child to comprehend everything fully and immediately.

Recent research indicates that children can understand the concept of quantity and even the symbolics of numbers earlier than the intricacies of the written word. They can distinguish between the size of two different piles of beans or pebbles long before they can make the abstract conclusion that language has spoken and written means of expression. Although they might not be able to count the pebbles before they are a few years old, they can distinguish between bigger and smaller, more and less, higher and lower.

Parents should include comparison and sorting activities in their children's daily life, using the appropriate vocabulary. For example, rather than call two different-sized sunflowers a small and a big flower, it is important to introduce the principle of comparison and describe one flower as *bigger* than the other. Simple as this example may seem, the concept of comparison is essential before children can start to count or perform any other task with numbers and quantities.

Examples of maths activities are everywhere, and most of the time there is no need to set up specific games to learn the basics. Even in places like the library, which families usually visit for language-related reasons, you will find hundreds of opportunities to do simple maths with your child. Count the shelves that hold children's books; count all the books with a green cover in one row; compare between the formats and sizes of the books; count the pictures on a page or sort a pile of books by colour.

And have a look in the numbers and maths section in your local library children's section. Literally hundreds of books offer an introduction to mathematics suitable for children of all ages. From story books about numbers to mind games and mathematical tricks, you will find something suitable for every child's particular interest. I have also included a few suggestions at the end of this book (see Further Reading).

At home, let your child pour juice into different-sized glasses and guess beforehand which glass would hold more juice. Or use containers with the same volume but different shapes to pour the same amount of liquid from one to the other, guessing whether it will all fit. Younger children might not be able to express their ideas about width, volume and size straight away, but they will have a familiar feeling about the concept when they come across a more formal introduction later.

Food preparation is another good source of maths-related activities. Many children are quite keen to help and if they need any encouragement at all, the prospect that the family or guests will soon have to eat a cake prepared by the family's youngest member will almost certainly help to motivate.

Preparing piles of different ingredients, watching an apple being cut into four parts or many slices, waiting for half an hour for the mixture to cool down, setting the oven dial – these are all experiences a child might not fully understand at the time but which still are helpful preparation for more co-ordinated instructions in maths later. Similar activities can be found in many books and are within the reach of every parent, no matter how adverse their own feelings against mathematics are.

Even the step from having an idea about quantity and comparison to transferring the idea to some written symbol is easier for children than transferring spoken words into written language. To transfer something you can think and say into mathematical symbols requires only one 'translation'. For example, to describe the fact that three beads were added to the pile that already had five beads, all they need to write down is 3 + 5. As John Glenn points out in his book *Towards a Numerate Society*, young children are capable of writing and understanding 3 + 5 much earlier than they can cope with the written phrase 'three plus five'. To record the same thing in words they need go through several levels of creating: first putting together letters that combine to words, and then combining the words to make a statement.

Undeniably, maths can become very complex and abstract at a more advanced level, but early and basic mathematics is a lot easier than the thinking processes required to understand how we write and read. However, as Glenn emphasises, it is important to train and strengthen both skills and try to avoid a gap between the two. Advancement too fast in one skill may act as a block to learning the other. The best training is to combine *all* skills in the early exercises. If your child enjoys working with beads you can get him or her to note down the mathematical symbols for any transaction, while you write down the words. Later, in another session, you can both read back from either the maths or language notebook and repeat the transaction with the piles of beads.

Numbers can be introduced as soon as a child can speak. The symbols for numbers are quite ubiquitous. They start with the child's own street number, telephone number and the phone dial, numbers on grocery packages, the clock, television screen or remote control . . . many other ordinary household items will offer plenty for a child looking for mathematical brain fodder.

If children could enter school knowing that maths is in fact everywhere, that it can be lots of fun, and with a vague idea about how basic mathematical operations like addition, subtraction and multiplication work, they would have an enormous advantage, not only in mathematics as such but also in their motivation to include maths in their repertoire of future learning and study. Just as children need to learn to distinguish between sounds and have an understanding of the symbolism of letters before they can learn to read at school, we can spare our children from having to learn maths without knowing about the basics or its daily use.

Arithmetic, the science of numbers, is only one area of mathematics and young children are likely to come across geometry, the science of shapes, lines and curves, even earlier than they start understanding numbers. Children know about the roundness of a ball long before they grasp the concept of three dimensions and the more abstract notion of the circle

Maths words

Counting: one, two, three, as well as 1, 2, 3 . . .
Comparison: bigger, smaller, more, less, fewer, longer, shorter . . .
Shapes: circle, rectangle, square, line, sphere . . .
Quality: round, long, curved, in spirals . . .
Maths operations: plus, minus, times, equals, divided by . . .

as the two-dimensional equivalent. They will take any opportunity to feel, see and otherwise experience all kinds of shapes and surfaces. Balls and cardboard boxes of various sizes can keep a young child busy for hours, trying to put the smaller boxes or balls inside the bigger ones or building towers with them. Cans provide an opportunity to find out which side they roll on and which they stand on. Learning mathematics requires the ability to recognise likenesses and differences and to sort objects accordingly. In order to be able to do this, children need a great deal of experience with the vocabulary that describes these relationships.

It is often the parents rather than their children who 'make a detour' around maths, thereby preparing the ground for later unease or frustration when the subject becomes more difficult. But even parents who feel totally inadequate about maths can give their children enough assistance with the basics and encourage them to find their answers for more complex questions elsewhere. Although maths might seem unimportant, difficult and threatening to you, you should always remember that your child is coming to this free of any such hangups.

Arithmetic and geometry should always be introduced with the appropriate terminology, incorporating language as well as numerical skills. The first book on number work could well be the first book of language, too.

Mathematics is like an universal language and its simple forms are more accessible for children than any other language. While we all recognise the problems associated with dyslexia and illiteracy, many children remain basically innumerate throughout school and struggle to gain even basic skills in adulthood. But the majority of jobs today are unthinkable without at least a basic understanding of maths; in fact even everyday things like reading the paper require some knowledge.

Literacy versus numeracy

Children will learn the spoken number words in the normal context of learning to speak. At this stage, they will use words like one or two, more or less, many and few, in context with something countable. For example, there may be many hens but only one dog.

When children start learning the symbols for letters they are just as capable of starting writing and recognising number symbols. These skills could be even easier to acquire because children are more likely to encounter individual numbers, while letters will more often be part of a word.

In *Towards a Numerate Society*, John Glenn says that children can abstract quite early from using number words in context with countable things to using the symbols 1, 2, 3, etc., independently. With the knowledge of only a few symbols, true arithmetic can begin. Most children don't need to verbalise or translate number symbols into language soon after they start using them.

Mathematics soon acquires the status of a separate language, and most of us would never consider writing 3 + 5 as 'three plus five'. However, for a young child it is important to practise as many methods of recording a mathematical process as possible – spoken words, mathematical symbols, actions with beads, fingers, etc., and written words.

Children might never again get the same amount of individual attention as in their pre-school years at home, and because the amount of information and its complexity will increase over the years, it will be harder to learn unfamiliar concepts later.

No matter whether children will later attend school or be home-taught, parents who have helped their children during the pre-school years will find it easier to gain their attention during later stages of teaching.

I often find it difficult to draw up some sort of teaching schedule for parents who ask me how to teach their young children maths. Each approach has to be adjusted to the child's developmental stage and its personality. But all activities should encourage children's choice and active participation. Parents should support their children to persevere with a task, get fully engaged for at least some of the time, take small risks and try things out, test the results themselves and regulate their own pace.

Active participation and practical exercises are particularly important for maths. Although many children will understand most principles in theory, actually writing down some exercises, repeating an exercise and applying a certain technique to different examples will deepen the understanding and help the child to memorise the particular operation. Mathematics is not a spectator sport and understanding can only be achieved by actually participating. None of this can be enforced on children, however. They have to dictate their hunger for learning and, within certain limits, they should be able to decide when to do what.

If children feel encouraged and safe, they are likely to develop a natural sense of competence and a strong desire for learning. In a supportive climate they will see learning experiences as adventures, be proud of their achievements, and develop an awareness of the fact that learning is not something we have to do but something that can be very enjoyable and is most successful when it happens by choice.

Maths sessions don't need to be long, but they should be regular. Parents should always back off as soon as the child loses interest. Only gross mistakes should be corrected as gently as possible but children should never be punished for making mistakes. Simply go back to examples the child already knows and keep practising these.

First steps
The home atmosphere has to make learning possible and easy. The child should have an area where they can work quietly and undisturbed, but not entirely isolated from the rest of the family. These conditions are important for children who are being prepared for school or supported during school as well as for those who will be entirely home-taught. We had a very informal and not particularly co-ordinated approach to our children's education during their first years. Michael, our youngest, started learning arithmetic at the age of four when we were sure that he could hold a pen and scribble on paper without difficulty.

Once I start with more formal sessions, my teaching has a strong focus on practice. Many of my students confirm that they not only understand maths much better by regular practice but also develop a better memory for principles or examples and begin to see the connections between various topics and rules.

Arithmetic is just like any other language in that children can first speak it and later learn the symbols to write the same expressions. As soon as a child knows how to count to 10 and what the numbers from 1 to 9 look like, simple arithmetic operations can begin. I believe most children are quite fascinated by counting and are happy to practise it whenever the opportunity arises. Again, ordinary household chores can be useful in combining an exercise in counting with an activity the child might want to participate in anyway. Counting the cups, plates or cutlery when children dry the dishes, or counting the pegs when they help hanging out the washing, are only a few of many possibilities.

To make the next step towards the link between the amount counted and its written symbol, children might need a lot of repetition, practice and encouragement. All methods applied when children are taught to read are just as helpful here. Memorising numbers is easier if you use cards with all the numbers as symbols as well as written words, have simple exercise sessions writing all the numbers down, or take a discovery trip through the grocery packages to find all the numbers and read them correctly.

From early on I would recommend having an exercise book for each child where they can scribble in their work and follow their own progress. But at the same time, don't put any pressure on your children. Just as a parent is expectant but patient enough to wait until a child utters its first word, it might take some time before children can write the symbols correctly and without confusing them. They are also likely to make many mistakes along the way; but this is an important part of the learning process, and although mistakes should be corrected this should not carry a 'you-got-it-wrong' message. Rather, correct mistakes by *showing the right way*. For example, one of my students, an eight-year-girl. wrote the figure 6 the wrong way round; instead of correcting her each time, I continued writing my 6 correctly until she finally picked it up herself.

Counting and the first arithmetic exercises will be easier if supported by anything countable that can be found in a household, like coloured pebbles or beans or a ruler. Once counting and writing of numbers comes naturally, the pebbles will also be helpful with the first attempts at adding and multiplying.

Some parents and teachers think that children should stop counting with their fingers as soon as possible, and actively discourage them. We once tried to help a seven-year-old girl who came to her teacher's attention because she was noticeably slower than other children. During the sessions with us it became obvious that she had never developed an abstract concept of numbers and still needed to count with the help of her fingers. However, she had been told that only small children needed to do that, so she tried to hide her hands and count her fingers without looking at them. She had problems with co-ordination and developed a behavioural problem that gradually grew into a real learning difficulty. Unfortunately, the sessions with us only improved her skills marginally but we could not really help her because she had

At the beginning . . .

Use anything you can to help the child count –
  fingers, beads, marbles, rulers, etc.
Say everything you do aloud: talk yourself and
  your child through the exercise
Write it down in mathematical symbols as well as
  words (if your child can't write yet you might
  want to do that anyway while they are watching, and you can keep the notebooks for later)
Don't correct mistakes – demonstrate the correct
  way

never had an opportunity to move on from counting things
to thinking of numbers in abstract terms.

Mathematical operations
As children start adding, they need to comprehend the structure of the denary system. Although they can distinguish
between all numbers from one to nine, they need to understand how these few symbols can be used to build up an
infinite range of numbers. This step towards the two-digit
number can be quite confusing, and anything to illustrate
the process will help. Several ten-centimetre paper strips
painted with the numbers 1 to 10, 11 to 20, 21 to 30 . . . or
piles of ten pebbles to count can help withunderstanding.
  It is essential that children have an idea about the denary
system even if they cannot count fluently to a hundred. As
soon as children know how to count and write one to nine
they will almost automatically use their fingers to arrive at
ten and regard ten as a sort of unit. The easiest explanation
of the denary system is that our counting system (as opposed
to the old Roman or Egyptian system) counts in lots of ten.
  One way to introduce the idea of counting in units is to
write down the number of days before your child's birthday,
Christmas or any other important event in the form of strokes
on a blackboard.

If you have to wait for eighteen days for an important event to happen, you could write them down like this:

IIIIIIIIIIIIIIIIII

This would be very hard to recognise for most people, and we would probably automatically write something like:

‌HHH HHH HHH III

Children might only just be able to count to five but they will still find it a lot easier to 'package' the strokes so that the large number becomes clearer. The denary system is based on the very similar principle of collecting in tens. Once ten 'packages' have been collected, they can again be piled together in hundreds, and so on.

We all make some sort of structured collections and there are many other ways to illustrate this for a child. For example, yoghurt often comes in six-packs. Such everyday examples make it easier for a child to get used to the various systems. But for most children, curiosity and motivation to count beyond ten will be the main driving force to learn the symbols and words for the numbers, and they will gradually develop an understanding of our counting system. Only much later will they fully understand the finer details. Do not be dogmatic about the way they learn – what matters is that they should enjoy the learning and arrive at the right answer however they do so.

One of the children I taught was quite good at adding even really large numbers and she had a good idea about the difference between tens, hundreds and thousands. But her teachers wanted her to do her additions according to the number's place value.

Instead of calculating her additions like this

$$
\begin{array}{r}
4362 \\
+\ 624 \\
\hline
4986
\end{array}
$$

she was expected to split each number into place values and
add them separately

$$4000 + 300 + 60 + 2$$
$$+ 600 + 20 + 4$$
$$\overline{4000 + 900 + 80 + 6}$$
$$= 4986$$

This exercise confused her more than it helped her to
understand place values, and had a negative effect on her
generally good addition skills.

Starting arithmetic

Addition
With young pre-school children, addition will develop from
adding simple one-digit examples (e.g. 2 + 1 = 3) with the help
of fingers or piles of pebbles. First examples of addition
should be as simple as possible so that the child has the
chance to take in the symbols + and = as well. It is important to talk through the examples, encouraging the child to
repeat the example aloud and to ask questions. Each of the
examples should be recorded in the exercise book – children
need plenty of practice before they begin to remember the
results by heart instead of having to count through each
example.

I would keep the practising to the one-digit level for as
long as it takes children to feel really comfortable with it.
The way to judge this is to invite them to tell you when they
want to move on, and go along with them. This encourages
autonomous learning.

With practice, most children will soon be able to remember additions up to 20, which is plenty for a pre-schooler
unless the child is interested and keen to be accelerated even
further. To make it easier for them to memorise the results,
additions can be compiled systematically:

To get to 2:	1 + 1 = 2
To get to 3:	1 + 2 = 3
	2 + 1 = 3 (and so on)
To get to 4:	1 + 3 = 4
	2 + 2 = 4
	3 + 1 = 4 (and so on)

At this stage, the symbol and value of zero can be omitted as it has not much meaning to a child and is too abstract for a beginner. But as soon as the mathematical operations extend to ten, then the zero fills out what would otherwise be an empty space. This is where understanding of the denary system becomes important so that children begin to understand that counting is really nothing else than a sequence of counts to ten.

Children will soon become familiar with this process and will be able to give results verbally and in writing without having to count through one by one. Practical sessions don't have to be school-like but can happen spontaneously at any time: while shopping, waiting for something, or during a walk along the beach. As often as possible, practice should combine both oral and written forms of record-keeping.

The more advanced examples of addition will gradually teach a child that the order of any two numbers added can be reversed without changing the result. Without stressing this rule too much (it is also known as the *commutative rule* for addition), it can still be pointed out to children at this early stage so that they will recognise it when it comes up again with multiplication.

The more examples children practise in which this rule is shown, the easier will it be for them to transfer it to later arithmetical operations.

The next step would be adding three instead of only two numbers, a progression leading gradually to multiplication.

$$1 + 2 + 3 = 6 \text{ or } 2 + 3 + 3 = 8$$

are examples even pre-school children will soon be able to answer.

Apart from advancing the child's addition skills, these examples also include further mathematical rules and principles children will encounter later. In particular, when dealing with sums of three numbers, children will learn that addition is always carried out with two numbers at a time (in the last example, $2 + 3 = 5$; then $5 + 4 = 9$). They also will discover that they can add a series of numbers in any order and still get the same answer (this is also known as the *associative rule* for addition):

$1 + 2 + 3 = 6$ can be solved by adding the first pair first and then the resulting sum with the third number or vice-versa.

Thus

$$(1 + 2) + 3 = 3 + 3 = 6$$

is the same as

$$1 + (2 + 3) = 1 + 5 = 6$$

and

$$(2 + 3) + 3 = 5 + 3 = 8$$

is the same as

$$2 + (3 + 3) = 2 + 6 = 8$$

Both the commutative and associative rules should be included – but you don't need to teach these terms or explain further the meaning of these principles. Just give them enough examples so it will be easier to recall the concept when the same principle is encountered later.

Multiplication

The above examples easily lead to simple multiplications as this operation is nothing more than a repeated addition.

$$3 + 3 = 6$$

can be read as three plus three, but children will often real- ise themselves that it is the same as taking three twice.

$$3 + 3$$

becomes two lots of three.

$$1 + 1 + 1 + 1 + 1 + 1$$

becomes five lots of one, and so on.

The multiplication symbol (x) and terms such as 'times' and 'is the product of' should be introduced gradually and at a pace the child can manage. The child will soon realise the advantage of expressing repetitive addition as multiplication: it is much quicker to write 5 x 2 than 2 + 2 + 2 + 2 + 2. This has the potential to be one of the delightful early discoveries they make.

At any time, if the child runs into difficulty or just risks losing interest because they are not quite ready for a particu- lar task, they should be referred back to the examples they can already do easily. Children will know best when they are ready for the next step and if things become too complicated the matter becomes frightening and confusing. Instead of trying to achieve understanding by repetition only, children should always be the ones to set the pace and know that there is no shame in going back to what they already know.

With the first forays into multiplication, many children will come across the first of many notorious 'sticking points' – the multiplication tables.

For beginners, the answers should not go beyond 18, which is the maximum for the addition if only one-digit num- bers are included in the practice. Further advances will depend on the individual child's hunger for mathematics, the

parents' or tutor's enthusiasm and general learning circum-
stances.

Michael was quite keen to do maths and has always
enjoyed the subject, but he also often told me that he needed
more time before tackling a particular topic or example.
Sometimes he only needed a short break and took off to play
sports. Often, he knew he simply could not understand the
example yet and he would keep it and come back to it later. If
Michael represents somebody who is interested in maths and
is highly motivated, children with other preferences might
need even more time away from their maths and progress
more slowly through the material.

Having tutored children who had strong adverse feelings
about the subject, I believe the single most important thing
is to make every effort to undo feelings of dislike. Children
should always know that it is up to them to set the pace of
progress, and parents should let them know that they appreciate their efforts and realise they are doing their best. As a
learner of entirely new material, a child is constantly performing at his or her boundaries, gradually expanding in
knowledge and understanding. Although much of what young
children learn may seem trivial to adults, we should never
forget that they are taking in enormous amounts of new
information every day.

I always tell children I am aware of the fact that they are
making an effort and that I am more than happy to help
whenever they want to move on. But it's up to them to set the
pace. I also don't go over the top with praise because it might
put children under pressure to fulfil further expectations. I
do, however, always try and make the children feel they have
achieved a lot already and that I will make it easier for them
if they want to go further. I tell them that what they already
know is plenty, no matter how they feel about it in their classroom and peer group, and that further steps forward will
become easier each time.

Multiplication tables often create the first hurdle for
children. Most children are perfectly capable of understanding multiplication as an operation with numbers but they
struggle to learn the tables by heart and fail to perform mul-

tiplications at the speed required in tests. I always have a colourful tables poster on the wall, but I never put students under pressure to learn the tables by heart. *They'll learn them anyway,* by saying and writing each exercise enough times. Pressure and high expectations will only make it harder. You can create a board game or other games to practise maths exercises but I have often found that children actually enjoy going through many relatively easy exercises if they can just set their own pace and rhythm.

Calculators

I don't start using the calculator in sessions before I'm reasonably confident children know most of their times tables, at least to the extent that they would pick up a wrong result. They will realise that it is even faster if they know their tables reasonably well. The general understanding of any mathematical operation is better the more a child can follow it through step by step. They should always be able to judge whether the calculator's results make sense, and pick up wrong results or miskeyed entries.

At this early stage of mathematical learning I would also encourage them to keep practising their multiplication until it comes easily, and to check their answers by using the tables on the wall or the calculator. However, in some areas of more advanced mathematics – for example, logarithms and trigonometry – calculators have almost totally replaced tables and hand methods. Insisting on these old methods would confuse children rather than help them with their understanding.

Subtraction

Subtraction can also create problems for many children. Although adding and subtracting are complementary processes, the concept of the latter seems to be more difficult for many. Subtraction requires a higher level of abstraction, but is very easily illustrated with the same tools used for addition. Some children are quite abstract thinkers and may not need much explaining and illustrating, while others will grasp the concept better if they think through everyday examples. If there are six bowls of dessert in the fridge for a family of four, how many bowls will you still have tomorrow? I am sure there is always a real-life example which will spark the interest of even the most unmotivated child!

When practising subtraction, it needs to be emphasised that it is the reverse of addition and that if

$$4 - 2 = 2$$

then similarly
$$2 + 2 = 4$$

Younger children may find it easier to understand the link between addition and subtraction if it is approached like this: Adding 5 to 4 gives 9, and if we take 5 away from 9 it gives us the 4 again.

It is also important to demonstrate the fact that numbers cannot be reversed in subtraction without changing the result. This is very different from the situation with addition and multiplication explained earlier. (It also holds with division, as we will shortly see). Some children will pick it up themselves, depending on how often they practised the rule when adding, others might need to try it out to see that the result really changes.

Thus, although

$$4 + 3 = 3 + 4$$
$$4 - 3 \neq 3 - 4$$

($\neq$ means 'does not equal')

At this point, you need to decide whether to introduce negative numbers or postpone such advanced material. With my own children, I moved through the levels of complexity much faster than they would have at school because they seemed interested and capable, but children vary and it might be better to teach them division first, before introducing negative numbers and the decimal point.

Many of the school children who come to me for maths tutoring have had problems that arose from misunderstandings or lack of practice with these basic operations. Because arithmetic operations build up from these basics, even small gaps can cause problems and contribute to increasing levels of frustration.

With children who are already having problems I always start with examples they find easy. First of all it is important for them to feel that they can actually do maths before they move on to the topics they find difficult. Knowing that you will help if they need it, they will then continue doing more examples that are similar enough to keep them succeeding, but which include new challenges to keep them motivated. The best motivation is the fact that they can solve the exercises. Soon they will develop the urge to go on to more complicated examples. Unless a student strikes a problem and asks for detailed explanation I let them learn through practice.

Division and fractions

One of the mathematical concepts most children dread is simple divisions and fractions. While additions and multiplications, and even subtractions, are still somehow more easy, divisions and fractions, and especially further mathematical operations with fractions, are often their first experience of rather abstract concepts.

However, the topic can be made to come alive by using the same principles and illustrations as with additions. Coloured beads, coins or anything countable will help to teach the concept of division. For example, twelve sweets can be placed in six groups of two, four groups of three, and so on. Children will enjoy this kind of practical example.

Aspects of division will find their way into a family home easily at dinner-time as tables are set, helpings served, dessert bowls distributed. Division takes place whenever anything is being shared out. For example, how many pieces of cake will you get if there are four of us and we have cut the cake in eight pieces? Examples like this can be drawn, or even cut out if that will help the child to understand.

Children should learn that division is complementary to multiplication, but the order of the numbers in the sum is important. The relationship between multiplication and

A summary of simple mathematical operations

✓ Use the right terminology, don't shy away from maths words.

✓ Correct mistakes by example.

✓ There's nothing wrong with repeating exercises the child already knows well.

✓ Tell the child repeatedly that you are aware of the effort they're making.

✓ Let the child set the pace.

✓ The denary system is like counting in units of tens over and over again.

✓ When you add, you can swap the numbers without changing the result, e.g. 2 + 3 is the same as 3 + 2. (This is *commutative*. Children don't need to learn this name, just the concept.)

✓ The same applies when you are adding a whole string of numbers – e.g. 1 + (2 + 3) is the same as (1 + 2) + 3. (This is *associative*.)

✓ Repetitive addition of the same number (e.g. 2 + 2 + 2) gradually leads on to multiplication.

✓ Multiplication is also commutative.

✓ Subtraction is the reverse of addition, and division is the reverse of multiplication.

✓ Neither subtraction nor division is commutative, i.e. 2 – 4 is not the same as 4 – 2, and 4 ÷ 2 is not the same as 2 ÷ 4.

✓ There can be different ways of writing mathematical exercises down.

division is similar to the one between addition and subtraction and the following connection is important:

$$3 \times 4 = 4 \times 3 = 12$$

Also:

$$12 \div 4 = 3$$

and

$$12 \div 3 = 4$$

but

$$12 \div 3 \neq 12 \div 4$$

Division is easy to illustrate or to practise at home, but should also be practised in writing.

Sometimes parents will come across a topic they still remember well from their school days, but they find a different method in their children's exercise books. Subtraction and long division are examples where methods have changed over the years. Often your child will be able to explain one of the examples, and they are likely to enjoy teaching their parent for a change. If you feel you still can't make much sense of the exercise, the textbook might offer a clue at the beginning of the chapter, or your child's maths teachers should be happy to explain the method to you.

In most cases the changes are mainly in the way the examples and the intermediate results are noted down.

There are various ways to write an example for long division:

$$8452 \div 4 = 2113$$

could be written as

$$\begin{array}{r} 2113 \\ 4\overline{)8452} \end{array}$$

In this method, each of the numbers is divided by 4 separately and the results are recorded on top. You start with 8 and divide it by 4 and note the result, 2, as the first digit. You apply the same process with all numbers: $4 \div 4 = 1$, $5 \div 4 = 1$ but you also keep 1, which means that instead of dividing 2 by 4 you have 12, which, divided by 4 equals 3.

The same example could also be written like this:

$$
\begin{array}{r}
2113 \\
4\overline{)8452} \\
8 \\
\overline{04} \\
4 \\
\overline{05} \\
4 \\
\overline{12} \\
12 \\
\overline{0}
\end{array}
$$

Here, too, each number is divided by 4, but all the intermediate results are noted, especially when numbers have to be retained.

Learning after school

THE MAJORITY OF SCHOOL-AGE CHILDREN ARE already used to learning sessions and can be taught in more structured modules ranging from a few minutes up to two hours. But it is important to create an atmosphere in which the learning becomes part of a normal day and not a chore.

Most parents will probably feel enthusiastic enough about teaching their children. In most cases it is more likely to be the subject that causes a motivation problem, but I can only repeat how important it is to leave any dislike and negative feelings about maths behind.

The discovery that maths can be quite fascinating, not as difficult and complicated as we have learned to believe, will come as you allow yourself to become involved. A minimum of enthusiasm for teaching and the subject is essential – not just for you to enjoy your work but also for the children to accept and respect you. Mutual respect can only develop if children sense that you are genuinely interested in imparting your knowledge and that you are not judging them on their performance.

Pressure and expectations will only add to the stresses and fears most children experience at school already and therefore the atmosphere with your child or a group of children at home should be as informal as possible.

However, children have to know that they will be working during whatever time has been allotted for the teaching session. Never mislead children and promise hours of games and fun when what you really want to do with them is their homework in maths or a few exercises from the textbook. The fun will come in the form of a sense of achievement and a feeling of reward for work completed. But children will find it easier to concentrate if they know this particular hour has been allotted for maths and will be followed by time for other activities.

At our home, I prefer working in groups rather than with individuals. It gives the children an opportunity to experience a social environment not too different from school, and they automatically keep practising important social skills such as overcoming the embarrassment of asking questions in a group, listening to explanations others receive in case they could be helpful, interacting with others, helping each other, having the courage to be as non-competitive as possible and being involved in general group dynamics. On average, I teach groups of four and move around among the children to help as problems arise.

With each new child I will spend the first of normally ten two-hour sessions establishing that my part in the process is to be there for them and to answer all questions. Their part will be to ask for help early enough – before they spend too long trying to solve a problem or getting the wrong result.

One mother dropped off her son for his first session promising he would work very hard for me. The boy didn't exactly give the impression he was looking forward to spending the next hours doing maths, and his mother's promise made him feel even more uncomfortable. He became noticeably more interested when I told his mother that there was no need for him to work hard. I said it was my turn to work hard and that he had come to discover he could enjoy maths.

I try to introduce new children into a group approximately reflecting their age, but this is not really important because during the session they will all move on at their own pace, cover different topics and develop an individual relationship with me. If you have more than one child, they can easily have learning sessions together, lthough each of them might be working on different exercises and at a different level of difficulty. Each child should work on their own material and at their own pace.

The right posture and a good working environment are basic but important. The desk should have an adjustable lamp and enough space to spread out their books and writing gear. Children should find their most comfortable position to work for a certain amount of time but they should always be able to take breaks and to stretch out a bit.

When I teach my students, I ask them to keep one finger on the exercise they are working on and to move through the material with both hand and brain. However, I am flexible with these things, because while some children leave their finger on the page automatically, others may feel uncomfortable with this. A woollen thread is used as a bookmark linking the working page with the answers page simply to make it easier to flick back and forth.

At the beginning of each session I start children off on one of their problem topics and ask them to do exercises with which they feel still confident. I help them with the first example, writing the exercise, including the question or task, into their exercise book with a little note about the book page and topic. Simultaneously I demonstrate what I do, step by step, emphasising the crucial steps. Before I leave them to start their own exercise and move on to start off the next student, I emphasise they are encouraged to ask me for help if they get stuck.

Although this means that the whole group will always be

exposed to background conversations I believe the one-to-one approach within a group is important and more effective for two reasons. It reinforces the personal relationship between me and each of my students, and I am sure that I can assist them with their individual problems so that they don't get bored or lose their attention. Yet, I prefer to develop a one-to-one relationship with each member of a group rather than teaching them individually in separate sessions. Apart from offering them a more social environment, I am also always aware of the risk of making them feel different. For many children, no matter whether they are considered to be gifted or whether they come for remedial maths, having private tutoring is a little embarrassing, and to see that there are others in the same situation makes it easier for them.

In smaller groups it is quite easy to keep close contact with each student, but even when teaching bigger classes I encourage each child to come up to me with their questions. Again, I use humour and analogies to help the students overcome their shyness or embarrassment. At one school where I teach a group of twenty-five young children, I play the role of the owner of a fish-and-chips shop who complains about a lack of customers, if none of the pupils seems to need me. These role games, appropriate for any age group, will help to encourage students to participate and ask questions.

Each child will have encountered a different problem, something they missed, did not understand or had a spell of low motivation. The answers to my inquiries about where the trouble lies vary from 'I don't understand fractions' to 'I hate maths' or even 'I hate school'.

Students who know what their problems are only need to rehearse the particular topic with someone who will explain and answer questions. Exercises of varying complexity will increase the level of competence. I try not to ask them to do too many exercises, to avoid boredom and loss of interest. Just before they feel they have done enough I try to move on either to the next topic or to a more complicated version of the same problem.

Some encouragement to keep going already comes with the right result for each of the exercises, but I try to add to that personal success by constantly recognising their effort. If they can't solve a problem and they ask me for help, I thank them because I am genuinely grateful that they have accepted my help and started identifying and tackling their problems.

Most children (and adults) are pleased to accept encouragement. They quite readily accept when I tell them they did not fail because they were not bright enough and that it will not take very long for them to be able to solve all the problems and even more difficult ones.

With a child who has difficulty identifying specific problem areas I would start with exercises appropriate for the age group and ask him or her to let me know if it is too hard. It might require a step-by-step explanation of a concept or exercise but even the most de-motivated students only need more time and more recognition to regain the general motivation to learn. From there they grow into self-starters and with each successful attempt at solving a problem they gain enormously in courage and speed.

In many classrooms, and unfortunately in many homes as well, the impact of a bad mark is much longer-lasting than that of a successful test. All too quickly school children are convinced that they can't do it and that they will never be able to achieve in a particular subject or even generally at school. I have to undo many of these convictions by an unbroken supply of encouragement, which often takes more time and energy on both sides than the actual subject, maths.

Of course, there are also the students who come to be extended beyond their normal school learning or to move through their subject faster, for example to be able to sit school certificate exams earlier. Most of these students have a high degree of motivation and are therefore not too difficult to teach. Yet, their own learning can be enhanced by teaching them skills and methods to take control, and encouraging them to transfer these skills to other subjects. I try to guide all my students from dependence to independence and from conformity to well-adjusted individuality.

As an important step to becoming a self-starter, an autonomous learner, all my students learn how to check their own results. Most maths books provide a section with results and

I will always give them the results with any exercise I might add to the selection in their books. They are always allowed to see the results when they approach an exercise.

Surprisingly, many students use these results only to determine whether their own solution was right or wrong but don't go beyond that. I ask them to stop their practising each time they have a differing result and go through the exercise with me. Together we find the missing step or wrong calculation, but instead of just telling them at which point they went wrong I try to phrase it in a more encouraging way: 'You were on the right track here, you just forgot this or that' or similar phrases which keep the students involved in the problem and will therefore encourage them to go back and look at it again.

I have found many of my students are easily motivated by the fact that they know the results already. They then tend to spend more time actually thinking about their exercise and trying to get it right. They learn to be in charge of their own learning, to define their own progress and pace, and to accept their own encouragement without waiting for somebody else's praise. This creates a learning environment where each student develops a rhythm of moving through the exercises, checking their results and ticking off each right solution in their exercise book. I ask them to do their own checking to subtly indicate to me to mind my own business: they are their own boss and can always call me if they need help. They know that there is someone who will help with exercises they cannot solve but who will not interfere otherwise.

It can be very heartwarming to see a child who was previously bored by maths, turning the pages to the results section after increasingly short intervals and with a proud smile.

All my students write with a pen instead of a pencil. Any mistakes they make should stay in their exercise books instead of being rubbed out. Although some are uncomfortable with this at the start, they soon realise I never regard mistakes as failures. I might have to explain the particular topic or concept several times but I never blame the student for mistakes or unsuccessful attempts at solutions. I blame myself as their teacher for their difficulties – and I do so quite openly, not to make them feel guilty or think they have disappointed me but to re-instill some of their self-esteem.

Sometimes I run the risk that they may become too confident, but their respect for me levels out most of their over-the-top reactions. After a few sessions my students realise that I respect them genuinely and that I regard them as equals. Soon the pleasure of their work, and its rewards, will come from the pleasure of learning and understanding itself and they will need less encouragement from me.

One of the difficulties in a family or class situation is that some sort of hierarchy will almost inevitably develop in most groups. The students in my classes will have established their 'pecking order' in other groups and are liable to do this again. To avoid this, I try to involve the students in as many interactive situations as possible. They are encouraged to help each other and even take on the teacher's part occasionally: to teach me how they see the best approach to a problem. None of this happens against the student's will. It is not meant to be a test of their competence or confidence. But my sessions are generally very informal and often situations like this arise on the spur of the moment.

Use encouraging language

- ✔ That's good
- ✔ You did that well.
- ✔ Now that you are doing that so well, can I show you something more?
- ✔ This is where it went wrong; you'll get it right next time.
- ✔ See? It's not so difficult, is it?
- ✔ Would you like to move on now?
- ✔ Would you like to do more, or try something new?
- ✔ You're getting so good now that you can move on to more advanced work if you like.

During all these small-scale and protected simulations of real-life group dynamics, the panic many children feel when confronted with new maths concepts disappears or at least diminishes. They begin to approach maths – and not only maths – with more confidence.

Henry Ford's saying, 'If you think you can, you can; if you think you can't, you can't,' might be considered a cliché now, but many of my students have confirmed his words. Whether they believe they can or can't do it, they will always be right. It is therefore imperative for me to instill some general self-esteem and confidence while I am teaching the actual subject.

It's often okay to let children see the answer to a problem before they do it. Knowing the form of the answer in advance often provides a clue, and provides encouragement.

Swap roles frequently and let the children teach you.

In a group-teaching situation, prevent hierarchies developing among children by encouraging co-operation: keep the emphasis on co-operation.

Encouragement and discipline

Parenting, teaching and learning is still often based on the principles of punishment and reward. Many parents use threats of punishment or withdrawal of love to make their children do what they want them to do. From a more positive angle, little bribes and promised reward often serve as incentives for children to please their parents. I believe these 'educational tools' have persisted for so long because they seem to achieve their goal. Children will stop making noise if they are threatened with grounding for the rest of the week; they will probably sit down and do their homework if they are promised a new toy or something they have wanted for some time.

But do children really learn much apart from blind obedience? I believe punishment is more about power and control than about bringing up children. Our children should mean more than that to us. Children who are bribed and rewarded don't have much of a chance to grow into independent and self-assured young people. They will always need to please somebody; they will need somebody's approval for their ideas and for their well-being they will need an unbroken supply of love and admiration.

Children who are punished or threatened with punishment will experience humiliation and intimidation. They may indeed stop doing what they were not supposed to do but all they will have learned is appropriate behaviour for a particular situation. It is likely they will not understand why a certain behaviour is appropriate, but only act in a way which avoids the punishment.

We have avoided both principles in our family. Of course we gave Michael a present when he passed his Bursary exams, but he got his cricket bat for passing the exam, not for passing it *earlier* than others. A big achievement like this certainly merits a reward, but on a smaller scale, we have avoided bribes, inducement, persuasion and pressure.

Instead we tried to encourage our children and to give them the feeling that we believed they could achieve whatever they had in mind, within reason. Encouragement is very inspiring and strengthening. It helps children to feel assured; to develop a sense of self-worth and independence in decision-making. If children feel loved, cared for, encouraged and supported, usually self-discipline comes with it without any extra effort. If they feel part of a family and know they are an important member surrounded by people who believe in them, there is no reason for accumulating aggression, letting out frustration or being malicious.

Even so, children will occasionally do things parents and caregivers don't want them to. We should make an effort to

> ✔ Self-discipline usually develops automatically in an environment of love, support and encouragement.
> ✔ Avoid punishment and negative messages.
> ✔ Regard mistakes as an opportunity to learn.

explain why we don't want such behaviour, rather than punish them in the hope they won't do it again. My main problem with punishment is that is always hurts twice. It might hurt physically or emotionally but it also inevitably destroys a child's dignity and prompts them to start building a protective wall. Even minor punishment will make it impossible for a child to regain full and unconditional trust.

More maths

Adding fractions

Things become more complicated when divisions are written as fractions and they become part of further calculations. To start teaching addition and subtraction of fractions I would include an example where the individual fractions are so simple that it becomes easier to calculate the fractions first. This helps to remind the students that the fraction is actually nothing else than the mathematical expression for dividing one quantity by another.

The terms numerator (the top line of the fraction – e.g. the 3 in the expression $^3/_4$) and denominator (the bottom line) can be explained using a simple example and should be introduced gradually into the terminology. For example:

$$\frac{12}{3} + \frac{30}{5} = ?$$

I always read the example aloud with the student: 'Twelve over three plus thirty over five. Twelve and thirty are the numerators and three and five are the denominators.'

In this case it can be easier to calculate the individual fraction and do the additions later:

$$\frac{12}{3} \div \frac{3}{3} = \frac{4}{1} = 4 \quad \text{and} \quad \frac{30}{5} \div \frac{5}{5} = \frac{6}{1} = 6$$

'Twelve over three divided by three over three equals four and thirty over five divided by five over five equals six.'

Then:

$$\frac{12}{3} + \frac{30}{5} = 4 + 6 = 10$$

'Twelve over three plus thirty over five equals four plus six, equals ten.'

Now that my student knows the answer, I can go through the general procedures of solving additions and subtractions involving fractions, which can be applied to all exercises. Again, I read aloud every step I write in the exercise book with the student.

$$\frac{1}{3} + \frac{2}{5} = ?$$

To add two fractions with different denominators both fractions have to be converted to a common denominator: a number that can be divided by both 3 and 5. (In this context, the word 'common' means 'the same'.) To find the common denominator, multiply the two denominators together: i.e. 3 x 5 or 5 x 3.

Here we see how the commutative rule for multiplication (see page 71) plays an important part in feeling comfortable when dealing with more complex operations. However, if you change the bottom line (denominator) of a fraction you have to change the top line (numerator) in the same way in order to keep the same value. In the example above, both the top and bottom lines are

$$\frac{1}{3} = \frac{1}{3} \times \frac{5}{5} = \frac{5}{15} \quad \text{and} \quad \frac{2}{5} = \frac{2}{5} \times \frac{3}{3} = \frac{6}{15}$$

and therefore

$$\frac{1}{3} + \frac{2}{5} = \frac{5}{15} + \frac{6}{15}$$

In other words, you multiply both lines of the first fraction with five (the denominator of the second fraction) and multiply both lines of the second fraction by three (the denominator of the first fraction).

Then the two fractions will have the same (common) denominator, so that they can be combined into one single fraction. Thus:

$$\frac{5 + 6}{15} = \frac{11}{15}$$

While writing down the exercise I would read it with the student step by step and explain further if necessary. Although some students might not ask any questions I sometimes notice that they are still uncomfortable with one of the steps and I will explain it in a sidebar.

It will be necessary to explain why the numerators are added, but not the denominators. For example, that the denominator shows how many pieces the cake is cut into, and the numerator shows how many pieces are eaten.

If a child struggles with these rules, you can always go back to more simple or graphical explanations. For example, consider a children's birthday party. At the end of the party are still several cakes left over so you give each child something to take home. One child wants to take home a quarter of a cake for herself and a quarter for Mum.

Mathematically, you would write that as:

$$\frac{1}{4} + \frac{1}{4}$$

The example will make clear that the child is taking home *two* quarters of cake, or half a cake, but not just *one* quarter which would be the result if you added the denominators.

$$\frac{1}{4} + \frac{1}{4} = \frac{2}{4} = \frac{1}{2}$$

Subtracting fractions
For subtraction I would choose another simple example to explain the process, pointing out with each step that it is exactly the same as for additions of fractions.

$$\frac{21}{25} - \frac{16}{25} = \frac{5}{25} = \frac{1}{5}$$

Solving the individual fractions first can again make the calculations easier and give the student confidence. But the same principle of finding a common denominator for the fractions applies for subtractions as well as for additions.

$$\frac{3}{4} - \frac{2}{3} = \frac{9}{12} - \frac{8}{12} = \frac{9-8}{12} = \frac{1}{12}$$

Note again that when you do the actual subtraction step, you don't subtract on the bottom line, only the top. To explain this, the cake example above can again be used in reverse: if you have two quarters of cake and take away one of them, what do you have?

How long a student would spend doing these examples depends entirely on how long it takes for them to learn the principle. I start with simple examples to give both options: to calculate the fractions first, or find the common denominator. The satisfaction of being able to prove your own result through another method is very motivating for many students and it gives them the confidence to trust the method later when the solution is not quite as obvious or, as in more complicated examples, the individual fractions can not be calculated separately.

Multiplying fractions

Multiplying fractions is easier in a way because all the student needs to do is to multiply both the top and bottom line, then check whether there is a common factor enabling them to reduce the fraction down to a simpler form.

For example:

$$\frac{9}{10} \times \frac{5}{12} = ?$$

There are two options: reduce the fractions first and then multiply:

$$\frac{\overset{3}{\cancel{9}}}{\underset{2}{\cancel{10}}} \times \frac{\overset{1}{\cancel{5}}}{\underset{4}{\cancel{12}}} = \frac{3}{8}$$

or multiply and then reduce:

$$\frac{9}{10} \times \frac{5}{12} = \frac{45}{120} = \frac{9}{24} = \frac{3}{8}$$

It would be difficult for a child to reduce the fraction $^{45}/_{120}$ in one step, so this can be done in two steps.

First, divide top and bottom by 5:

$$\frac{45}{120} \div \frac{5}{5} = \frac{9}{24}$$

then divide by 3:

$$\frac{9}{24} \div \frac{3}{3} = \frac{3}{8}$$

By showing the student both methods, it can be pointed out that the first is easier and that he or she should always look to 'cancelling' the fractions first before multiplying.

Who needs fractions? Everyone does!

Your child might later live in a flat and it could be his or her job to calculate the power bill, grocery spendings or share in rent.

To buy a new appliance on hire purchase, the shop requires a certain fraction of the price as a deposit.

The adult dose of a medicine is 20 ml. How much would your baby need if it should have $^{1}/_{8}$th of the adult dose?

A recipe gives the quantities to make eight waffles. What fraction of the recipe should you make if you want only six waffles?

The maths teacher has told you that your son has been talking for three-quarters of the lesson. If the lesson is 50 minutes long, how long has he been talking?

Dividing fractions

Dividing fractions is almost as simple as multiplying them: you can turn a division into a multiplication simply by 'flipping' the numerator and denominator of the second fraction (ie. the one doing the dividing).

For example:

$$\frac{2}{3} \div \frac{4}{9} = ?$$

By inverting the second fraction the original division problem becomes a multiplication:

$$\frac{2}{3} \times \frac{9}{4}$$

Now all the rules of multiplication of fractions apply again. Encourage the student to cancel before multiplying:

$$\frac{\overset{1}{2}}{\underset{1}{3}} \times \frac{\overset{3}{9}}{\underset{2}{4}} = \frac{3}{2} = 1\frac{1}{2}$$

The concept of mixed fractions can be illustrated by $^3/_2$. Show that out of three halves, you can make one whole cake and have half left.

Thus

$$\frac{3}{2} = 1\frac{1}{2}$$

A common problem experienced by many students is distinguishing between fractions like $^2/_3$ and $^3/_2$. Giving them the choice of which fraction of a chocolate bar they would choose and letting them keep the spoils of that choice soon teaches them the difference!

Introducing decimals

The introduction of decimals should go hand in hand with learning fractions, because decimals are just another way of writing fractions. The only difference is that, with decimals, the fractions are expressed out of either 10, 100, 1000 or 10,000, etc.

Children are quick to follow the rule that you 'put a point for the one, and a figure for every nought after the one'.

Lots of practice with sums such as

$$\frac{1}{10} = 0.1 \qquad \frac{1}{100} = 0.01 \qquad \frac{1}{1000} = 0.001$$

$$\frac{3}{10} = 0.3 \qquad \frac{3}{100} = 0.03 \qquad \frac{3}{1000} = 0.003$$

$$\frac{17}{10} = 1\frac{7}{10} = 1.7 \qquad \frac{17}{100} = 0.17 \qquad \frac{17}{1000} = 0.017$$

will make this concept graphically clear.

Once that is grasped, the next step can be introduced. What do we do with fractions like $\frac{1}{2}$, $\frac{1}{5}$, $\frac{1}{4}$ and $\frac{1}{20}$?

Remind them of the rule that the denominator of the fraction must be either 10, 100 or 1000, etc., so the fraction has to be converted. Thus

$$\frac{1}{2} \times \frac{5}{5} = \frac{5}{10} = 0.5$$

and

$$\frac{1}{4} \times \frac{25}{25} = \frac{25}{100} = 0.25$$

School could be the best party in town

MANY SCHOOLS ARE GROWING TO REALISE THAT the one-to-one or a small-group approach to teaching is far more effective and enjoyable for teacher and learner alike, and that children will learn better the more they are in charge of their own progress. Teaching has been changing from lecturing and testing children to coaching them to become self-starters and autonomous learners. Special programmes and so-called individualised education plans are popping up everywhere but these initiatives are generally designed for children who already stand out. The programmes are targeting the few children who show obvious signs of excellence and are underachieving in the mainstream classroom either because of a lack of opportunity or because they feel it is socially more acceptable to be mediocre.

I would argue that we should offer more individualised programmes to every child, no matter how outstanding their achievements are or how wide their range of interests is. We should allow for the whole range of personalities to develop, whether the young child is a promising athlete who has obvious abilities in many other school subjects and even an interest in the arts, or a badly demotivated child struggling with most school subjects and at risk of getting into serious trouble.

It is easy to see why we should help those who struggle but I think all children would benefit from a teaching environment where they could set their own pace and determine how much energy they will invest into any of the subjects and activities. Of course there have to be guidelines and certain limits, but I believe that the schools' output on an academic and any other scale could improve with a higher degree of self-determination for the children.

Many teachers and parents argue that it would be impossible to run a school under these principles: that the schools' limited resources only enable them to offer acceleration programmes to the obviously gifted. But the children who come to my attention are mainly children at each end of a performance scale. They either draw attention because they excel in at least one subject or they stand out because their behaviour is unacceptable.

The majority of children would probably pass through school without any opportunity to check how far they could go if they were given control over their learning. Those whom we 'diagnose' as gifted stand out because their special achievement happens to be in an area that we regard as worthwhile. I am convinced all children would have the capacity to go further and achieve many goals earlier than they are doing now if we would approach them with a non-judgemental and positive attitude. And if we allowed all children to progress at their own pace, teaching could even be less stressful and exhausting for the teachers, too.

The greatest problem, of course, is 'the system', not the people in it. But despite the difficulties, everybody has the responsibility – and privilege – to be creative and innovative, to improve themselves and the system.

Glen Stenhouse writes in *Confident Children* that the development of a strong sense of self-worth, competence and trust is the result of many opportunities to make choices, see them through and then evaluate the results in order to have a guide for future choices.

He says schools could encourage the growth of self-reliance and independence by giving students an effective say in aspects of running a school, greater input into the structure of their own learning programmes, and by involving them in the assessment of their work and behaviour.

One of my students came to do maths with me to be able to pass School Certificate as early as possible and to leave school for a music academy. As a fourteen-year-old, she had a vague idea that she wanted to be an opera singer, and her interest in maths was understandably limited, but the simple fact that she rediscovered how to enjoy learning not only helped her to perform better in maths but also made her even more determined to put all her energy into her chosen career. If her teachers had been more supportive, she would have been the easiest pupil to teach.

She was quite happy to accept that she had to fulfil at least a minimum standard in a range of general subjects but if allowed to follow her interests as much as she wanted her motivation to do the other subjects would have been much greater. But her potential in a very specialised area had neither been recognised nor acknowledged, which made her very hard to teach and very frustrated. She had a lot of initiative and a high level of motivation and curiosity, but felt held back and hindered in her progress most of the time. At some stage she tried to start a band at her school but, although many other pupils were interested in joining, she felt there was no support within the school.

Of course these feelings had an effect on all her learning, not only on her progress in her chosen subject, music. She felt there was no incentive to work hard because her school never acknowledged any of her ideas, initiatives or achievements. However, when she changed schools and made a fresh start with new teachers who she felt recognised her particular skills and interests, she became a very popular and top student.

Children's natural motivation is like a huge pool of energy, which we all too often waste or fail to make the best use of. If children felt more involved in the decision-making, if they knew that they would find support for their interests, they would find it a lot easier to accept having to spend some time with subjects they might not find particularly interesting. I believe school could be a lot more fun for both teacher and pupils if we relaxed some of the rules and goals and let each child move through at their own pace, with a focus on core subjects like maths, language and learning skills, and with their own choice of additional subjects. School could and should be the best place in town for children to spend their time.

Some of these ideas are being implemented at the moment. A qualification framework is being introduced in New Zealand's secondary schools, based on the idea that each students should have the opportunity to follow any particular interests at their own pace and be credited individually for their progress, not by comparison with other students in the same year. Some of these interests could exceed what the school offers in any particular year and the new framework allows students to move on to more advanced levels. Other interests might be extra-curricular and perhaps not be included in the school's repertoire at all. Students will be allowed to attend classes at other institutions or enrol in correspondence courses, with credit for their achievement being included in their overall qualification.

The Ministry of Education proposes a 'seamless education', allowing for more flexibility and individual choice, based on the idea that active choice almost guarantees a high level of motivation. Although this would place new requirements on the flexibility of teachers, it would bring the learner into the centre of the process. The maths curriculum has also been changing towards a more individualised exploration and discovery oriented approach of teaching.

In and out of school?

THE LEVEL OF FREEDOM TO CHOOSE COULD EVEN go as far as deciding whether to go to school at all. I would like to see a school system working like a revolving door. We would still have the legal obligation to make sure our children receive education, but they would be free to move between school, home-schooling and correspondence courses at their own or their parents' initiative. New Zealand has many children enrolled in correspondence school already and the qualifications achieved through correspondence are acknowledged and transferable to any other form of schooling. Correspondence school began in the 1920s as a way of educating children who lived in remote areas. Today, parents who want to home-school can draw on resources and expertise from the correspondence school system. An additional benefit is that parents probably improve their own skills in many subjects (as well as in teaching practice) as soon as they start teaching their children.

Rather than having to decide for one or the other, home schooling and school should be supplementary and complementary, with the common goal to give children the best possible access to education. While children at home might have a better chance to develop their own interests, this home advantage should go hand-in-hand with the social and inter-active skills best acquired at school.

We had taught David, our eldest son, at home although he also attended primary school. Because of his additional teaching in maths, he soon surpassed his classmates and could have moved on to university level in maths. He would have been happy to continue with all his other subjects and attend maths lectures at university or by correspondence, but it was a major bureaucratic battle to get permission for him to go to university at his age. It was only with the sup-port of a number of teachers and university lecturers that David was finally allowed to go to university. Both David and later Audrey were required to go through a psychological test to confirm that they were mature enough to cope with their own progress.

Things had become a lot easier by the time Michael faced the same problem, but we still encountered a lot of suspi-cion and we found ourselves having to justify why Michael should want to move through the subject matter so quickly. Yet, in 1994 we happily accepted an invitation to be one of New Zealand's ambassador families during the Year of the Family and we felt proud to be selected because we provided an example of a family learning together.

Methods and approaches are changing in education, and I believe all this is to the better. Yet many people will still ask whether we should extend and accelerate our children in the first place – just because it is possible to do so? With my acceleration method, which gives each child the choice and opportunity to take in as much as they want and can at any moment and with a strong focus on learning skills, I believe I can give children an advantage that goes far beyond better comprehension of any particular subject. It prepares them for many other challenges.

Learning doesn't finish when we leave school or univer-sity, and we need to equip our children with the necessary skills. Teaching children the subject matter as well as quali-ties such as helpfulness, modesty and feelings of self-worth and confidence will make it easier for them to tackle any future challenge and to keep developing new skills through-out their life.

My forté is to incorporate all these qualities into my teaching and to show children that they can in fact be high achievers, have fun doing so and not be arrogant about it. As long as we offer acceleration to only a few selected children who happen to perform well under current conditions, we run the risk of creating an even more competitive learning environment.

Imagine how a child feels after missing out on selection for acceleration classes: the obvious message is that the child is 'not good enough': not worth supporting and accelerating. If children are constantly put down and intimidated, they will take this experience into their adult lives and perhaps never achieve much at all. But why shouldn't we all be tall poppies?

Everybody who has succeeded at coaching and encour-aging children who have lost their belief in themselves will agree that children are always worth our unconditional sup-port.

Mathematics should be one of the base-building subjects and started as early as possible. The number of children who are basically innumerate is alarming at a time when technol-ogy is so quickly advancing and most people need at least a basic understanding of maths in any field. Even in this 'computer age', there are far more innumerate than illiterate people.

Most effort to help these people focuses on literacy. How-ever, I believe a solid grounding in maths is just as important as language skills and much easier to teach early. Pre-school children and new entrants should be taught maths and lan-guage skills in combination with learning skills from the start and gradually be introduced to other subjects to give them an opportunity to test out what they enjoy. Children who have a good knowledge of maths will be able to branch out into any subject or area that interests them, with a better chance of success.

Teachers and family as a team

ALTHOUGH MOST PARENTS WOULD SAY THEY WANT the best education for their children, not all are confident enough to teach their children. Maths is no exception and many children learn quite early not to expect much help, if any, from their parents. We too sometimes had difficulty helping our children but instead of simply telling them to get help elsewhere we were honest enough to say that we did not know the answer.

Family support and a close and co-operative relationship between parents and teachers is very important for creating an atmosphere conducive to learning. Teachers and parents together should be the prime support team for every child. Together with the child, they should try and work out the best educational plan. However, I have seen cases where teacher and parent almost considered each other as rivals.

Teachers sometimes feel as if parents are infringing their area of expertise and parents often complain that teachers don't support their child enough or not in an appropriate way. Many teachers regard parents' participation in education as interference with their job, and parents are often unhappy when their children don't get as much attention as they would want.

I think we all need to make an effort to appreciate and value both parenting and teaching more. They are two of the most difficult jobs. Parents are best qualified to support, recognise and advance their children's skills and abilities, and teachers have the knowledge and pedagogic skills to add on to that – so why should they not work together?

In mathematics, the teacher's qualifications and personal attitudes can often be a problem and many children might fail in maths partly because of that. A bad teacher can often destroy a child's interest in mathematics, particularly in the junior years when it is easy to put a child off maths for good. Such a person should however, not be treated harshly but encouraged and helped to find other work.

The ideal school

IN MY IDEAL SCHOOL, CHILDREN WOULD BE FREE to move between subjects and difficulty levels at their own pace. There would be no homework and no tests. Age would be used only as an initial guideline, but children would move through schooling at their own pace, so age differences become largely irrelevant. Without exams and with a flexible age structure, the learning environment would be much less competitive.

The only structured requirement would be that projects in specific subjects would have to be completed by a certain date. Teachers would be very approachable and children encouraged to develop learning techniques which would gradually make them autonomous in their learning. Mistakes would be regarded as an important part of learning and there would be no punishment or telling off. Children would be in charge of their own learning, they would get a lot of pleasure from learning and develop a sense of achievement and competence. They would know that it is not a waste of time to help somebody who still struggles to understand. They would help because they knew that they learned a new skill and experienced a new sense of pride whenever they explained something to another child. They would know that gaining all these other skills would be acknowledged as well.

These are all principles that I try to incorporate into my teaching. Although the subject is mainly mathematics, all these skills are essential for any student who wants to continue learning after they have finished with their sessions.

Just a few examples

ONE STUDENT WHO HAD RECENTLY EMIGRATED TO New Zealand was struggling to fit into the local school and education system. He had plans to study electronics and knew he needed a solid base in maths to do that. But maths had never been much more than a necessary evil for him.

'I decided to contact Mr Tan after I was told my standard of maths was probably not as good as that of a New Zealand school leaver and therefore not adequate to start studying at the polytechnic. My family had moved to New Zealand when I was sixteen and I had joined the sixth form at one of the local public high schools. However, during that year I focused on English and computer science. I had special tutoring by a teacher of English as a foreign language.

'I wanted to study electronics and knew I would need a solid base of knowledge in maths to do that. But I had never enjoyed maths, which made me rather reluctant to commit myself to spending any extra time on that subject. Maths was somewhat of a necessary evil and I was only prepared to put in minimum effort. At the best of times, I could get myself to feel indifferent towards the subject.

'When I enrolled with Mr Tan I was still not sure how much I would actually benefit from the sessions. I had finished my sixth form and wanted to begin with my studies at polytechnic with the new academic year. So I could only fit the extra tutoring in the summer break, which I spent working full-time for a landscape architect.

'Because of my summer job I had to schedule most of the sessions in evenings or Saturday mornings and given my dislike of maths my motivation was certainly not particularly high. The image I had about private tutoring did not help to motivate me either. I was expecting a strict teacher, constantly watching over my shoulder, and telling me off for mistakes.

'However, the first thing I learned was that I would have no problem with maths in future and that I would probably soon be ahead of what I would be expected to know. After asking where my difficulties were, Mr Tan started me on some exercises which I could not immediately solve and he just quickly showed me how to do it. The explanation came at the pace of a machine gun but it had nothing frightening about it. Somehow, although he spoke very fast, he managed to do so without putting any pressure on me. I just knew if I failed to understand he would be happy to explain again.

'Once I was confident to work on this particular type of exercise he gave me several similar problems to solve. The exercises were of the same pattern but included little variations and different approaches. Mr Tan started the whole class, mostly four to five students, on some topic. Each student would be working on an entirely different problem, either something they didn't understand in school or something they wanted to prepare for a test. I was infected by the pace with which Mr Tan spoke and moved around all of us and I began to go through my exercises one after the other. The atmosphere was that of a friendly competition.

'I practised so many exercises in relatively short time I was sure I could do them in my sleep. From one exercise to the next the process came more naturally and automatically. Although I was almost frantically working away at my list of exercises I felt increasingly relaxed and began to enjoy it. Something which I would have considered utterly boring became a challenge – and one I was confident to take on.

'Mr Tan would only interfere when I asked for help with something I could not solve. He was always happy to help and in his fast but encouraging and patient way, he would write down each step, simultaneously talking me through it and always checking that I was still with him. I was filling pages full of exercises and losing any track of time, yet I had fun doing it.

'In contrast to my experience at school, and later at polytechnic, I learned everything about the particular topics chosen for the session during the two hours spent with Mr Tan. At school, I always had to rework exercises or do some additional homework, but during these sessions I knew I had covered the topic fully and, even more importantly, understood what I was doing and what the fastest way of tackling the exercises was. Despite the fact that I never did any homework or prepared sessions with Mr Tan, the processes and methods have obviously been engraved into my long-term memory. Even weeks and months later I have no trouble solving exercises in areas of maths I worked through with him.

'Apart from the obvious improvement these few sessions had another positive side-effect. Previously, I would have looked at a mathematical problem, declared it as too difficult and not even tried to solve it. Anything to do with maths was just too daunting to approach. Now, although I often can't see any obvious or quick solutions, I know that I will find a way.

'My attitude towards mathematics, and in fact towards learning generally, has changed under the influence of Mr Tan. He gave me the motivation to enjoy learning, the confidence to approach new problems without help and a method of learning which helps me to retain things I have learned.'

I have also taught larger groups of children in schools. One school asked me to design a maths programme to accelerate primary school children in mathematics and here is what one of the school board trustees said.

'Choon Tan taught maths at our school for three terms when we started the primary school. We had opened a foundation school, a pre-school based on Montessori principles, three years before and decided to expand into the primary-school ages. But we wanted to offer the children the widest possible range of opportunities and decided to look for somebody outside the group of Montessori teachers.

'He taught a group of eight-to-nine-year-olds and trained a parent teacher and a maths graduate who were to continue the maths programme later. Soon after he began teaching, more and more younger children joined the group because their interest in maths had been inspired by their older friends. We ended up with a group of five-to-ten-year-old children voluntarily doing basic algebra and equations.

'They all enjoyed it and even the youngest coped well with exercises many other school children would struggle with. I am convinced that if you present a child with maths, in fact any subject, without inducing fear or giving warnings about difficulties, even very young children are extremely capable of learning. And Mr Tan did exactly that – he never made any negative comments, neither about his subject nor about any of his students. His students' achievements confirmed this approach.

'Mr Tan believes in each child's potential and he tells them about it. He constantly encourages those who might be reluctant or lacking confidence until they believe in their own potential. He presents the material in a positive way and he would never limit a person. He is always supportive and encourages any attempt at intellectual or creative activity.

'One of the reasons why we asked Mr Tan to teach maths at our school was because we wanted our students to be able to tackle subjects at their own speed. Our classes are not restricted by age and most of them would be a mixture of all primary school ages. Mr Tan approached each child individually and was happy to teach a class where older and younger students were mixed together.

'Often, the benefit for the children was even greater the younger they were. They were inspired by the older children and learned early that maths could be fun and quite rewarding. The younger children did not even start to question or doubt their own capability. They never had the negative attitude towards maths which is so often found with older children and adults.

'All the teachers were enthusiastic about having Mr Tan as a maths teacher and staff trainer. He was convinced each child was capable of achieving beyond any expected level in maths, even if it was not the child's favourite subject. The only difference between the children was the pace at which they moved through the material. But he never used this to compare them with each other. His individual approach helped to provide each child with enough support to make them feel positive about their own work and not compare themselves with other students.

'Mr Tan's method of teaching allowed each child to be an individual and to learn for themselves. All children responded very positively to the learning environment he created. They were having fun because they achieved something for themselves, sparked by their own motivation. Later, they were following their own enthusiasm which they developed for maths after a succession of successful attempts at it.

'The drive was to do better than last time; not to do better than your friend. Because competitive behaviour towards others was basically non-existent it did not influence the children's social relationships with each other. Mr Tan never labelled any of the children. He regarded them all as achievers because they were moving through the material presented to them and that was what mattered. There was no good or bad, quick or slow, smart or stupid.

'The children themselves adopted that approach and envy or jealousy were rare, although each group included some very outspoken and forthright as well as shy and reserved children. Mr Tan always positively reinforced the concept of autonomous learning and the children thrived in an atmosphere where it was up to them how much they did. They knew nobody would force them to do a certain number of exercises; but at the same time nobody would stop them if they wanted to do more. And there was all the help they needed if they could not solve a particular problem.

'Even making a mistake, something that can be very off-putting if it is emphasised too strongly, was no problem at all. The children knew they were not expected to get everything right and that they could ask for help. Mr Tan never judged any of the children, no matter how they performed. A mistake was never treated as a failure and, while correcting the error, he always found a lot of encouragement and praise for the child.

'Mr Tan managed to create an environment for the children in which they regarded him as one of them rather than a teacher in a position of authority. They respected him more for his friendship than for his knowledge, which made it possible for Mr Tan to share what he knew. His methods formed the basis of our maths programme.

'We have also extended Mr Tan's methods and techniques to make them applicable in pre-schools. Special equipment introducing the basic principles of maths such as the value of numbers, addition, subtraction and multiplication is available to children as young as two but only at their own initiative. Teachers are always available to play with children, allowing them to explain principles in a non-intrusive way. At primary level, eight-year-olds are now working on form five material – happily.'

A mother of a child with severe behavioural problems owing to a lack of intellectual inspiration and simply boredom approached me to set up an individualised programme for him. Her son was described as gifted, extremely able and advanced but also hyperactive with a severe attention-deficiency disorder.

'My son was what many would describe as a hyperactive child. He was dynamite. At the age of ten months he found out how to crawl out of his cot, and he was taking everything apart, examining things. Only a few months later his favourite occupation would be pointing at everything in the house followed by the question "What's this?"

'Our doctor told us that he was bright when he had a convulsion at four months of age. He said it was too early for children to have convulsions and that he had to be quite advanced in his development. His grandmother also observed that whenever he picked up a book he would hold it the right way. We ignored these early hints but when his kindy teacher told us that he was correcting her spelling we realised that he needed help.

'We lived out of town and he attended a rural school where the teachers insisted on starting him on junior reading material. They said he might miss something. But his frustration, and with this his behavioural problems, increased so we changed the school. He skipped one class and for a short while developed some interest in what was presented to him. But the relief lasted only briefly and he was soon very unhappy again.

'Because he considered it unnecessary to answer questions asked by a teacher who would know the answer anyway, he was often regarded as either slightly retarded or, as in most cases, socially deficient. This attitude did not help to make him feel happier or more motivated to collaborate or show interest.

'We sold our business and moved into town where he could attend a school which would allow some extension and acceleration. But at the age of eleven he had hardly been stimulated or challenged, and was still very immature and frustrated – it had become almost impossible to communicate with him.

'We contacted Mr Tan because maths was one of the most obvious subject where our son was unhappy. He had already been doing form five maths by correspondence, but it always felt like all the extending and acceleration had to be done almost secretly. Teachers discouraged him; peers envied him.

'Mr Tan simply took him as he was and accepted that he was doing very advanced maths. He was neither surprised by it nor particularly excited about it. It felt like it was the most ordinary thing to do – which was something our son probably experienced for the first time.

'He had finally found somebody who would let him go at his own speed and level and be genuinely supportive. Mr Tan decided to work on bursary maths with our son, which provided him with another first-time experience – he was challenged, and he made his first mistake. Looking back, this might have been the most important thing our son learned with Mr Tan. Making mistakes, and learning that it doesn't matter and that you only get it right when you get it wrong at first.

'Choon Tan was the first teacher to recognise and appreciate his skills and allow him to work at his own level and progress from there. Our son developed a genuine interest in intellectual activity, in challenge, had more in depth knowledge first in maths and later in other subjects, and, perhaps most importantly, had fun at doing it and began having more confidence in himself.

'Despite suggestions from others we decided against even faster acceleration and specialisation in only one subject. We wanted our son to stay with his peer group, and admittedly, for a long time we did not want him to be different; we wanted him to fit in. But it was obvious he related better to adults than to his peers and the out-of-school extension had a beneficial influence on his well-being and behaviour.

'So we began to build up a network of teachers for a variety of subjects or areas of interest. We offered everything we could think of to our son and whenever he was interested we sought out a good teacher who would give him individual tutoring.

'He continued the sessions with Mr Tan for several years, not regularly but quite frequently, learned to play two instruments, started new languages and got involved in sports. He also developed a growing interest in computer programming, which was sparked when he joined me in a basic computer course at the age of seven. This group of tutors helped to keep him challenged, busy and motivated, and made it possible for him to stay at school and still turn into a well-adjusted young person.

'Although many other teachers were involved, Mr Tan was the person who had set the course for this change for our son. He went from totally frustrated to happy and relaxed. By taking our son as he was without remarks about his abilities Mr Tan allowed him to accept himself and stop feeling this strange mixture of being bored and ashamed. And by taking him right through to his limits he made him make mistakes, which was a real breakthrough for all of us. This helped him to learn how to learn and he could transfer this skill to everything he approached since then.

"His emotional development also progressed very positively. He turned from a human doing to a human being. Now, at sixteen, he has a strong interest and skill in teaching others and is far more socially interactive than ever before, also with students of his own age.

'He has adopted a philosophy of OK-ness for himself and he began recognising skills in others. He still relates better to adults than to his peers, but his entire personality has changed to the better and he became much more likeable.

'Mr Tan is very non-judgemental, and being taught by him would be a very unique experience for anybody, no matter how capable they were – he creates an atmosphere where positive confidence is bound to develop.'

Resources, contacts and networks

For a start, your own kitchen, the supermarket, the beach or any other location your children enjoy can provide some inspiration to do maths. Often it is more a matter of realising that we take many things for granted which might be new, interesting and quite challenging for a young child.

When you walk past a parked car with your child, you might comment on its colour, size, etc., but many parents don't think of mentioning the fact that each car has four wheels, two at each side or at the front and back. It is these basic facts that we take for granted but which include a lot of basic mathematics, that can keep a child's mind busy during the first years, learning numerical skills.

Books expressly for children

Many children's books that introduce numbers and mathematical ideas comprehensively and creatively are available from any local library or bookseller. Our children particularly enjoyed Shirley Hughes' children's books but there are many more.

Their all-time favourite was *The Very Hungry Caterpillar* by Eric Carle, which first appeared as a book (originally published by World Publishing, New York, in 1969 and then by Hamish Hamilton, London, in 1986) and was later recorded as a talking book and video (BMG Video, Australia, 1994).

One Lonely Kakapo by Sandra Morris (Hodder & Stoughton, Auckland, 1991) is a lovely book which introduces the symbols from 1 to 10 through some of the author's watercolour paintings of New Zealand's birds and wildlife.

Based on a similar principle is *Sea Squares* by Joy Hulme and Carol Schwarz (Hyperion Books for Children, New York, 1991). Paintings of sea creatures accompanied by rhymes are combined to produce a lovely story book with numbers.

Look Around! A book about shapes by Leonard Everett Fisher (Viking Kestrel, New York, 1987) is for the very young but is also quite inspirational for parents who find it difficult to see the geometry and maths hidden in ordinary things.

Maths Is Childsplay by Jan Morrow (Longman, Harlow, UK, 1989) introduces a wide range of maths games and learning activities for young children. This book is full of ideas for play activities which involve numbers, counting or making comparisons.

A bit more advanced is *Mathematical Games and Puzzles* by Trevor Rice (Betsford, London, 1973). This collection includes many examples of optical illusions and mathematical board games.

Peggy Kaye's Games for Maths (Pantheon Books, New York, 1987) is another collection to keep your child interested in maths. The book includes examples ranging from pre-school to primary-school level.

Morris Minus and the Calculators by Tom Price (Viking, London, 1991) is an excellent combination of story telling and numerical skills, and introduces some basic mathematical terminology. Words like minus, plus, half, multiple, etc., are introduced in the context of an exciting story about and for children.

My First MathsBook by David and Wendy Clemson (R. D. Press, Sydney, 1994) introduces measures, weights, graphs, puzzles and examples of basic mathematical operations.

Take Me to Your Liter by Charles Keller and Gregory Filling (Pippin Press, New York, 1991) is a compilation of maths and science jokes, offering plenty of inspiration for teachers looking for something to fill maths sessions with humour.

The Story of Mathematics by Alistair Ross (Black, London, 1984) can be equally interesting for children and parents. It tells the story of how people began to measure time; about some of the famous Greek mathematicians like Pythagoras, and about counting systems used by other cultures.

Books for adults
Confident Children: developing your child's self-esteem by Glen Stenhouse (Oxford University Press, Auckland, Melbourne,

Oxford, New York, 1994).

The Third R: towards a numerate society, edited by John Glenn (Harper & Row, London, New York, 1978).

Preparing Young Children for Maths, a book of games by Claudia Zaslavsky (Schocken Books, New York, 1979).

Sharing Maths Learning with Children: a guide for parents, teachers and others by Pat Costello, Marj Horne, John Munro (Hawthorn Victoria, Australian Council for Educational Research, 1991).

Maths and Me: helping your child with mathematics by Avelyn Davidson Shortland Publications, Auckland, 1983).

Sense and Nonsense about Hothousing Children: a practical guide for parents and teachers by Michael Howe (BPS Books, Leicester, UK, 1990).

Many modern textbooks explain mathematics quite comprehensively; my favourites are *It's a Mathematical World*, Books 1 to 3, by W. Ellwood, P. Guerin, P. Purser and N. Rush (Longman Paul, Auckland), which have been revised and updated regularly since their first publication in 1987.

A new series of booklets by Wellington maths teacher Vivienne Petris has been published at the end of 1995. *Understanding Mathematics the Petris Way* is a series of ten booklets, each focusing on only one main topic such as trigonometry or quadratic equations. The booklets have been designed as workbooks in the real sense of the word: the author has left space to write and scribble exercises and solutions. Published by Heinemann, the booklets are targeted at fourth and fifth form, but could also be very useful for any parent who might feel uncomfortable about maths.

Waikato University publishes a regular journal called *Mathsplus* which includes many examples of maths activities and many essays about teaching and learning maths.

Educational software with a focus on maths is also available for those who enjoy computer-aided learning. *Treasure MathStorm* and *Math Blaster*, by the Learning Company, introduce basic mathematical operations and encourage practice as each of the tasks of the game depends on the player's mathematical competence.

Contacts and networks

Mathematical societies have also been formed in many centres or through individual schools. They are often a good source of material and can be helpful when motivation is the biggest problem. They organise many competitions and regular events and offer possibilities to start networks between parents, teachers and students — or to join existing networks. The groups' goal is to encourage interest in mathematics and to extend and promote mathematical scholarship.

The national organisation is:

New Zealand Association of
Maths Teachers (NZAMT)
PO Box 54151
Bucklands Beach
AUCKLAND

Contacts in regional groups are:

John Hutton
President/Secretary
Northland Maths Association
c/- Pompallier College
PO Box 10 042
WHANGAREI

Colin Anderson
Secretary
Auckland Maths Association
c/- Howick College
PO Box 38142
AUCKLAND

Melody Linton
Secretary
Bay of Plenty Maths Association
c/- Te Puke High School
PO Box 344
TE PUKE

Helen Newrick
Secretary
Waikato Maths Association
c/- Fairfield College
PO Box 12-228
HAMILTON

Adele Brooker
Secretary
Taranaki Maths Association
c/- Spotswood College
PO Box 6116
NEW PLYMOUTH

Jim Fountain
Secretary
Wanganui Maths Association
c/- Wanganui High School
PO Box 4022
WANGANUI

Ann Lawrence
Secretary, Manawatu Maths
Association
c/- Awatapu College
434 Botanical Road
PALMERSTON NORTH

John Scudder
Secretary
Hawke's Bay Maths Association
c/- Lindisfarne College
PO Box 2341
NAPIER

Andrew Ferguson
Secretary
Wairarapa Maths Association
c/- Wairarapa College
PO Box 463
MASTERTON

Di Claughton
Secretary
Wellington Maths Association
PO Box 12423
WELLINGTON

Lyndsay Pope
Secretary
Nelson-Marlborough Maths
Association
c/- Nelson College
Private Bag 16
NELSON

Rhona Lever
Secretary
Canterbury Maths Association
PO Box 31-014
CHRISTCHURCH

Rob Smith
Secretary
Aoraki Maths Association
c/- Waimate High School
Paul Street
WAIMATE

Anne-Marie Hutton
Secretary
Otago Maths Association
c/- Kings High School
270 Bayview Road
DUNEDIN

Ann Robinson
Secretary
Southland Maths Association
PO Box 97
INVERCARGILL

Similar interest groups have
formed for home-schooling
families:

The National Association of
Home Tutor Schemes
PO Box 12144
WELLINGTON

Ph (04) 473-6623
Fax 499-4943

Home-schooling is often
supported through material
from correspondence courses
and several of these courses
are offered throughout the
country targeted at children's
or adult levels of understanding.

Most correspondence schools
offer various courses to
Bursary level.

Correspondence School
Private Bag
WELLINGTON
Ph 04 473-6841
Fax 04 471-2406
(This includes the adult open
learning service.)

International Correspondence
School
PO Box 6026
Te Aro
WELLINGTON
Ph 04 384-4855
Fax 04 384-7381

Scott's Correspondence College
PO Box 30-990
LOWER HUTT
Ph 04 567-6592
Fax 04 569-3378

Open Polytechnic of New
Zealand
Wyndrum Ave
Private Bag 31914
LOWER HUTT
Ph 04 566-6189
Fax 4 566-5633

Further organisations offering practical help, material or information include:

Early Childhood Training Organisation
PO Box 894
PALMERSTON NORTH
Ph 06 357-4537
Fax 06 358-8635

Alpha Employment Training Ltd
PO Box 33 1155
Takapuna
AUCKLAND
Ph/fax 09 468-1052

Kiwi Superscope Training Institute
PO Box 311
PAPAKURA
Ph 09 297-7445
Fax 09 297-8095

Skills Update Training Institute
PO Box 51160, Pakuranga
AUCKLAND
Ph 09 827-3303
Fax 09 827-5855

Centre for Learning
PO Box 1729
CHRISTCHURCH
Ph 03 379-3980
Fax 03 366-2232

Avalon Education Centre
P O Box 31444
Lower Hutt
WELLINGTON
Ph 04 619-0600
Fax 04 576-4411.

Rural Education Activities Programmes are funded by the Ministry of Education and offer schemes tailored to meet the needs of rural communities. The programme varies from year to year.
Contact phone: (04) 417-6182

ALSANZ, the Accelerated Learning Society of Aotearoa New Zealand, brings out a regular magazine. PO Box 160, Raglan. Ph/fax 07 834-8811

Brainworks, a group promoting excellence in teaching and accelerated learning, publishes a regular magazine, organises meetings and sells educational material. PO Box 30 387, Lower Hutt, ph/fax 04 569-1112

Two directories offer an overview of courses and training opportunities available:

A Guide to Training Providers in New Zealand (published by Inteltech, PO Box 68 272, Nelson).

Excellence, New Zealand Education Directory (published annually by Suedden & Cervin Publishing Ltd, PO Box 68450, Newton, Auckland).

Index